The Tao of Cosmos

"This fascinating book is essential reading for both scientists and mystics alike! Ma skillfully integrates Western philosophy, Eastern wisdom, and modern science, offering a tour de force in his synthesis of a holographic approach to a new model of reality that unifies heaven, earth, and humanity. It stands as a magnificent contribution to the quest for that holy grail known as the Theory of Everything. His integration and skillful weaving of a coherent theory from a vast spectrum of knowledge across disparate fields is truly remarkable. Supported by meticulously detailed and captivating diagrams, charts, and advanced mathematics, this book describes his model in terms of a torus topology, integrating the classical yin/yang motif with elements of the I Ching, Leibniz's binary code, and contemporary DNA research discoveries.

Ma encourages readers to contemplate the boundless mysteries that go far beyond our current understanding of spacetime, urging us to consider humanity's place in a vast multiverse of multiple dimensions. A must-read for thinkers seeking to comprehend our role in the cosmological reality."

Shelli Renée Joye, Ph.D., author of *The Electromagnetic Brain* and *Tantric Psychophysics*

"This book details two key advances in science and cosmology and articulates the coupling between recent scientific achievements and the ancient paradigm of the Eastern neo-Confucian philosophy of the oneness of heaven and humanity. The author integrates Daoist ontology with Confucian cosmology to suggest a numero-cosmological system. He scientifically validates the unity of nature and man, showing that humanity is inextricably and holistically blended in the cosmological evolution.

This original cosmology—by a qualified scientist with a background in both astrophysics and Chinese philosophy—has the makings of a longstanding primary source."

Richard Grossinger, author of *Bottoming Out the Universe* and *Dreamtimes and Thoughtforms*

The Tao of Cosmos

The Holographic Unity of Heaven, Earth, and Humankind

A Sacred Planet Book

Zhen G. Ma, Ph.D.

Park Street Press
Rochester, Vermont

Park Street Press
One Park Street
Rochester, Vermont 05767
www.ParkStPress.com

Park Street Press is a division of Inner Traditions International

Sacred Planet Books are curated by Richard Grossinger, Inner Traditions editorial board member and cofounder and former publisher of North Atlantic Books. The Sacred Planet collection, published under the umbrella of the Inner Traditions family of imprints, includes works on the themes of consciousness, cosmology, alternative medicine, dreams, climate, permaculture, alchemy, shamanic studies, oracles, astrology, crystals, hyperobjects, locutions, and subtle bodies.

Portions of this book are based on the author's 2019 Ph.D. dissertation, "New Philosophy of Onto-Cosmology: Swimmean Paradigm Synthesizing Daoist Metaphysics and Big-Bang Physics."

Cataloging-in-Publication Data for this title is available from the Library of Congress

ISBN 978-1-64411-776-7 (print)
ISBN 978-1-64411-777-4 (ebook)

Printed and bound in the United States by Versa Press, Inc.

10 9 8 7 6 5 4 3 2 1

Text design by Virginia Scott Bowman and layout by Debbie Glogover
This book was typeset in Garamond Premier Pro with Degular, Gill Sans MT Pro, ITC Avant Garde Gothic Std, and Simsun used as display typefaces

To send correspondence to the author of this book, mail a first-class letter to the author c/o Inner Traditions • Bear & Company, One Park Street, Rochester, VT 05767, and we will forward the communication.

In Appreciation of the Beauty and Magnificence of the Universe

Contents

FOREWORD

Toward the Great Unity

Brian Thomas Swimme, Ph.D.

It is impossible to know what the future holds, but I have confidence in my prediction that the book you hold in your hands will be read by scholars a thousand years from now. To justify my confidence in making this judgment concerning *The Tao of Cosmos: The Holographic Unity of Heaven, Earth, and Humankind* by Zhen G. Ma, Ph.D., I will begin by placing his work in the history of cosmology to show its significance in shaping humanity.

In terms of Darwinian evolution, the first human ancestors emerged six or seven million years ago in Africa. For our purposes, the primary difference between humankind and primate species and chimpanzees is the eventual development of symbolic language by humans. Though the origin of symbolic communication is lost in the depths of time, some biological anthropologists place its beginning at the same time as the use of fire, some 1.8 million years ago. It is probable that for as long as humans have existed, we have pondered our place in the universe. *Cosmology* as a human activity appeared in different groups throughout Africa and then around the planet, where a *cosmologist* is understood to be someone who constructs a story concerning the origin and development of both the universe as a whole and the human species in particular. Many stories were articulated. Each of these stories established the fundamental values and meanings for each group. Celebrations of their shared story created bonds among groups of humans. Cultural anthropologists estimate tens of thousands of these cosmological stories appeared throughout the long history of *Homo sapiens*. Each of these stories was cherished so deeply, wars between different stories became common right up to the modern age. To take one example, in the sixteenth century, Protestants and Catholics began killing each other over differences between their cosmologies with a combined death toll estimated to be between six and nineteen million.

In the seventeenth century, a profound change took place. Humans discovered the scientific exploration of the universe which led to a new kind of cosmology, one that is empirically based, one that speaks not in terms of gods or titans or semi-divine animals creating the universe but of mathematical, physical laws that control evolution. Scientists from every civilization became involved with the development of this cosmology. After three centuries of work, the physical outlines of a universal creation story have emerged. In a single sentence this is the story of the universe that began fourteen billion years ago with the emergence of elementary particles in the form of primordial plasma, which quickly morphed into atoms of hydrogen, helium, and lithium; a hundred million years later, galaxies began to appear, and in one of these, the Milky Way, minerals arranged themselves into living cells that constructed advanced life, including evergreen trees, coral reefs, and the vertebrate nervous systems that humans used to discover this entire sequence of universe development. That sentence required four and a half centuries of scientific investigation. It is a crowning achievement of the scientific method.

Having established the history of cosmology as our context, we turn now to Ma's work. His bold assertion is that the empirically based, scientific story has a universal interpretation, a spiritual meaning, for all humans. That is to say, Ma has provided a synthesis of two vast domains of thought: the science of an empirically based account of evolution and the profound insights into cosmological meanings found in Daoism and Buddhism, spiritual systems that have been developed over thousands of years. The erudition demanded for this synthesis is vast. Ma has devoted decades of study in order to understand the intricacies and far-reaching consequences of contemporary mathematical cosmology. His study of Daoism and Buddhism has required an entire lifetime. Growing up in a family with a distinguished lineage of Daoist healers, he has been a student of this complex Asian practice and philosophy from his earliest childhood. *The Tao of Cosmos: The Holographic Unity of Heaven, Earth, and Humankind* is the world's first vision that combines Western mathematical cosmology with Asian wisdom at a high level of professional expertise.

One final comment. As you will see, Ma has drawn from my work on what I have called the "powers of the universe." I want to quickly indicate the major scientists who have discovered these powers. None of these powers is my invention. Each one comes from our centuries-long investigation of matter.

Power-1 Seamlessness refers to the discovery of the quantum vacuum. The physicist Paul Dirac was the first to propose the existence of this realm, which gives birth to elementary particles.

Power-2 Centration refers to the way the universe is composed of individual entities or centers. Among other scientists, the chemist John Dalton established that each thing in the universe is a composition of particles.

Power-3 Allurement is the attraction that binds entities together. The physicists Isaac Newton and James Clerk Maxwell constructed theories of two attractive forces, gravity and electromagnetism, that act throughout the universe.

Power-4 Emergence occurs throughout the entire history of evolution. The first scientist to articulate evolution at a cosmic scale was the mathematical cosmologist Georges Lemaître.

Power-5 Synergy—the power of relationships to birth something new—is at the base of biologist Lynn Margulis's work on the emergence of the eukaryotic cell.

Power-6 Transmutation—the power for organisms to adjust over time in order to fit into a community—was discovered by the biologist Charles Darwin.

Power-7 Transformation—the power to construct through evolution a more harmonious community—was proposed by the paleontologist Pierre Teilhard de Chardin.

Power-8 Cataclysm, leading to mass extinctions in life systems, was first elucidated by the zoologist Georges Cuvier.

Power-9 Homeostasis—the self-regulation of a system—was investigated in a deep way by the physiologist Claude Bernard.

Power-10 Interrelatedness was discovered by the botanist, naturalist, and geographer Alexander von Humboldt.

Power-11 Radiance is rooted in the Second Law of Thermodynamics as developed by the physicist Rudolf Clausius.

Power-12 Collapse and rebirth as an ongoing power in the universe is fundamental to the theories of the mathematical cosmologist Lee Smolin.

Ma has joined the ancient tradition of cosmology beginning in the distant past and pointing the way to a future of harmonious convergence. Where the first cosmologists constructed narratives that united groups of a couple dozen humans, and where later cosmologists proposed stories that united many millions in each of the major civilizations, Ma has taken the next step. With a profound ability for navigating both science and philosophy, he has bequeathed to the world a vision of the universe that draws from both Asia and the West. His work is a challenge to read. But in the mathematics and story is the glimmer of a new wisdom, a new understanding, a new beginning in humanity's long quest to establish a vibrant community that includes all humans and every sentient being.

Brian Thomas Swimme, Ph.D., a specialist in evolutionary cosmology, was educated at Santa Clara University and the University of Oregon. At the present time he is the director of the Center for the Story of the Universe and Professor Emeritus at the California Institute of Integral Studies in San Francisco. His most recent book is *Cosmogenesis: An Unveiling of the Expanding Universe.* Discover more about his work at storyoftheuniverse.org.

Acknowledgments

This book is a revised version of my third Ph.D. dissertation (2019), combined with additional illustrations and tables from my presentation at the fifth meeting of the Foundations of Mind (2017). The first and the last chapters are fully rewritten.

My first Ph.D. supervisor, Tsin-chi Yang, a prominent professor in high-voltage engineering in China (also a lifelong taiji practitioner) at Tsinghua University, Beijing, introduced me to Einstein's quest for a theory of everything (ToE) during the early 1990s. Eventually, in the late 2010s, I had the privilege of being under the guidance of Brian Thomas Swimme, a mathematical cosmologist, in the philosophy of cosmology and consciousness at the California Institute of Integral Studies (CIIS), San Francisco. It was through his mentorship that I discovered the path toward a quantitative ToE by integrating Eastern wisdom, Western evolutionism, and black-hole physics. I extend my heartfelt gratitude to him for his insightful ideas, critical feedback, and invaluable suggestions, which have greatly influenced my academic journey.

In addition, I would like to extend my gratitude to Professor Goodman, an expert in Asian philosophies and cultures from CIIS, and Professor J. F. Marc des Jardins, a Canadian Sinologist from Concordia University, Montreal. Both of them served as members of my Ph.D. committee. They provided invaluable advice during the project proposal and thesis completion stages. Particularly noteworthy is Professor Marc des Jardins's considerable time spent reading, reviewing, and providing feedback on the thesis to ensure the quality of the content. Undoubtedly, this book greatly benefits from his assistance, and I deeply appreciate it.

Furthermore, I am deeply indebted to Dr. Yi Wu, a professor emeritus and distinguished adjunct instructor from CIIS. He bestowed upon me a precious gift: an outdated bilingual version of the I Ching by Z. D. Sung (1892–1980), published in 1935. It was indeed one of the leading references in this book. Dr. Wu had diligently annotated numerous pages, offering insightful interpretations of the hexagrams and unveiling the hidden philosophical wisdom of Eastern numerocosmology. His notes proved invaluable in my research journey.

A special appreciation is extended to the Avatamsaka Buddhist Lotus Society (ABLS) in Milpitas, and the City of Ten-Thousand Buddhas in Ukiah. The ABLS provided partial financial assistance, while the City of Ten-Thousand Buddhas generously granted access to its library. Ven. Master Tian-Cheng at ABLS, an expert on the Avatamsaka Sūtra, provided invaluable insights into several Buddhist concepts from the sūtra.

Additionally, I express my sincere gratitude to Dr. Cuiwei Yang from Concordia University, Montreal, an expert in Chinese history and politics. Dr. Yang patiently offered detailed introductions and accurate comments on the formation and development of neo-Confucianism, the significant contributions of missionaries to Sino-West exchanges, and Shao Yong's neo-Confucian ideology.

In particular, I extend my heartfelt gratitude to the editorial department of my publisher, Inner Traditions. Transitioning from a weighty, scholarly Ph.D. dissertation to a commercially appealing book is an immensely challenging and distinct process, particularly within a short editing period. This task is compounded by the inclusion of various unconventional materials, such as Chinese texts, equations, extensive tables, and complex multicolor figures. These elements add layers of complexity to the editing and publishing process, requiring meticulous attention to detail and creative solutions to ensure the final product meets both academic standards and market expectations.

Considering all these factors, I am especially grateful to Jeanie Levitan, vice president and editor in chief, who offered detailed guidance regarding timing and scheduling. In addition, Courtney B. Jenkins, in her role, graciously extended the manuscript submission deadline, allowing for thorough proofreading. I am also very thankful for the kindness of all the staff, particularly the typesetters who perfectly designed and laid out this book. For example, the book cover appears mysterious, offering a farsighted view of our universe, and intriguing, capturing attention with its shocking imagery.

Among all, I am profoundly thankful to my editor, Kayla Toher, whose dedication and expertise have been instrumental in shaping the manuscript into its final form. Kayla's meticulous attention to detail, deep understanding of language nuances, and her patient, thorough, and professional approach has not only enhanced the readability of the text but has also elevated its overall quality. Moreover, Kayla's insightful feedback and suggestions and her willingness to go above and beyond in reviewing and refining the manuscript, always with patience and efficiency, speak volumes about her commitment to excellence. I am truly grateful for Kayla's unwavering support and her pivotal role in bringing this project to fruition.

Finally, I extend profound appreciation to Richard Grossinger, a distinguished American writer and the visionary founder of North Atlantic Books. His unwavering encouragement, insightful guidance, and steadfast support have been instrumental throughout the journey of bringing this book to fruition. Richard's deep insights into the literary world, coupled with his invaluable recommendations, have played a pivotal role in ensuring the timely publication of this work. His enduring inspiration continues to fuel my creative endeavors, and his mentorship serves as a beacon of wisdom and encouragement on this writing odyssey. I am truly grateful for Richard's belief in the potential of this project and for his enduring friendship, which has enriched both my professional and personal life beyond measure.

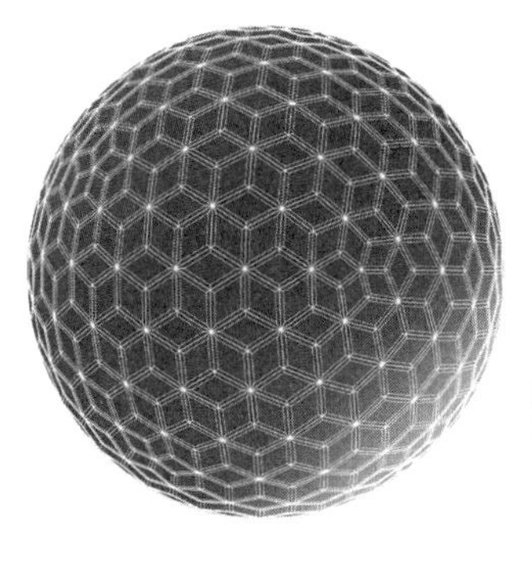

Introduction to the Holographic Unity of Heaven, Earth, and Humankind

During the scientific revolution of the seventeenth century, Sir Isaac Newton discovered the mathematical principles of natural philosophy. His work set up a solid foundation for classical physics. Human beings were therefore able to change the way they looked at and understood the universe. About two hundred years later, Albert Einstein developed the special and general theories of relativity, as well as the formula for mass-energy equivalence. He reformulated and extended Newton's laws in view of an absolute 3-dimensional (3-D) space and 1-dimensional (1-D) time to a curved 4-dimensional (4-D) space-time. The fundamental rules thus established were testable in explaining new astronomical observations. They became reliable in proposing novel and transformative theories that govern the entities of the reality we inhabit.

In the 1990s, Stephen Hawking and Sir Roger Penrose integrated quantum theory and the general theory of relativity. They proposed that the universe originates from quantum-gravity fluctuations. Penrose later suggested that the cosmological evolution follows infinite, self-sustaining, and conformal cycles. Afterward, he and Stuart Hameroff applied the quantum-gravity model to formulate a cosmic brain. They drew a connection between cosmic fine-scale structures and the neural microtubules in human brains. They suggested that the beat frequencies of brain waves show up in electroencephalography (EEG) as solitary packets. They concluded that, qualitatively, consciousness plays an intrinsic role in the universe. No doubt, such a novel scientific model expressed what has been a long-standing notion found in ancient Eastern wisdom, the oneness of heaven and humanity, the foundational theme in Daoist ontology and neo-Confucian cosmology.

This book introduces a quantitative theory of everything (ToE). It sheds light on recognizing the shared nature of human beings and the universe in which we humans find ourselves. Yet the theory appears to be a gamble from several perspectives.

First of all, I am a scientist in space physics and black hole research. Unusually, I have extended my expertise into the field of the philosophy of cosmology and consciousness. Supervised by mathematical cosmologist Brian Swimme, I engaged in integrating Penrose's modern cyclic cosmology, the ancient philosophy of the I Ching (or Yijing, known as the Book of Changes), and Swimme's spiritual evolutionism rooted in Christocentric cosmology.

In addition, Eastern philosophy indeed influenced Western natural and religious epistemology. For example, Gottfried Leibniz invented and expounded the binary arithmetic of number 0 and number 1 for computing on the basis of assimilating the numero-cosmological binary system of the I Ching, consisting of *yin* (denoted by number 0, referring to qualities that are, for example, feminine, negative, receptive, passive, dark, intuitive, still, grounded, reflective, restful, cool, spatial, and so on) and *yang* (denoted by number 1, referring to qualities that are, for example, masculine, positive, assertive, active, bright, logical, dynamic, expansive, projective, energetic, warm, temporal, and so on). However, still unresolved is how to reconcile the Western cosmic Christification and ancient Eastern atheism.

Furthermore, a spiritual evolutionism proposes a scientific onto-cosmological story, that is, the humanization of the cosmos. Such an approach was developed in the work of Teilhard de Chardin and Thomas Berry, both of whom postulated a highly progressive cosmic divinization. Swimme got rid of the religious overtones and articulated universal cosmogenetic principles. Yet it is still challenging to set up an atheistic evolutionism that is nevertheless based on an extremely developed cosmo-theology.

This book started as a thesis to cap my work at the California Institute of Integral Studies and has evolved to its present form. When first written, I painstakingly included Chinese characters and pinyin (the Chinese phonetic alphabet) to describe the concepts discussed in English. These can now be found at the back of this book in the Compendium of Celebrities and Chinese Terms and Texts (p. 188). There, interested readers will find a detailed list, in English and Chinese, of relevant Western and Chinese people with their lifespans, as well as translated terms, ideas, and verses cross-referenced with their original Chinese characters and pinyin. I hope the list's inclusion will serve as a thorough resource as you investigate how these concepts connect throughout my attempt to synthesize scientific, philosophical, and religious understandings and, quantitatively rather than qualitatively, account for a new onto-cosmological evolutionism of nature.

The book suggests a three-tier paradigm of cosmo-ecology to describe a cyclically and ceaselessly evolving universe. It articulates the unity of heaven, earth, and humanity by demonstrating a holographic principle for different evolving scales. The proposed evolutionism is in conformity with modern big-bang cosmology and the most advanced observations (e.g., the Planck Mission). What is more, the book introduces a toy model to explain Penrose's cyclic cosmology and formulates a nonlinear quantum-plasma dynamic to validate the Penrose-Hameroff cosmic-brain model via simulations of EEG waves. Throughout the book, a theme is demonstrated: humanity is inextricably and holistically blended in cosmological evolution and is able to unfold the mysteries of the universe through the flow of consciousness into mental activities.

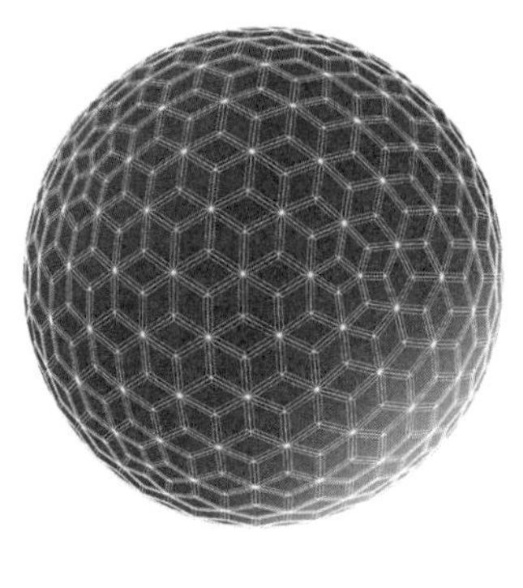

1
Statement of the Problem

EINSTEIN'S QUEST FOR A THEORY OF EVERYTHING

A theory of everything (ToE) aims at either providing an integrated paradigm of the universal and ultimate laws and truths about nature in view of "mathematical abstractions, experimental data, and physical intuitions" (Weinberg 1992, p. 5), or offering a conceptually integral vision of the world that qualitatively encompasses all the physical and spiritual dimensions, such as "matter, body, mind, soul, and spirit, as they appear in self, culture, and nature" (Wilber 2001, p. xii). Such efforts will eventually come to a close by accomplishing a cosmic universalism, that is, the holistic embrace of realities where all are recognized as intrinsically interdependent and/or interconnected, though extrinsically divergent when far away from each other (such as matter versus dark matter, or body versus mind).

Over the last few centuries, modern physics gained a rapid and dazzling expansion at the frontiers of scientific knowledge (Weinberg 1992, p. 3). The growth, in turn, left enough room for scientists to contemplate the scientific laws of the *uni*verse, investigate the cosmic features of the unknown *multi*verse, and predict new trends for the development of science. In view of modern cosmology, the *uni*verse was shown to be an integration of a variety of fundamental activities (e.g., electromagnetic interactions), laws (e.g., the second law of thermodynamics), and energies (or those in the form of powers, e.g., electricity, as well as forces, e.g., gravity). They were identified within scientific discourses as being responsible for all the complex processes throughout the evolution of the cosmos. From this perspective, human consciousness (or mind), an evolved holistic blending of the processes appearing at the end of the last 13.8 billion years, should be endowed with a "hominized" form of cosmic evolution, one that entails that human intellectual awakening (i.e., intuition or imagination, a priori insight, pure thought) is intimately aligned with the creative dynamics of nature (Swimme 2017a).

Before Einstein, most physicists, for example, Galileo, Newton, and James Maxwell, as well as their predecessors, followed the logical approach of ancient Greek positivism to construct scientific explanations of the objective reality. For them, empiricism took the leading role in the scientific method in which experience of observations, experiments, and comparisons was crucial to deriving basic concepts and laws of nature (Einstein 1932, p. 23). According to Auguste Comte, this was the last of the three successive phases in the quest for the truth of nature, those phases being the theological, the metaphysical, and the positivist (Giddens 1974, p. 1).

By contrast, intuition became crucial in the development of modern physics, resting as it did upon the quantum theory proposed by Max Planck in 1900, and both the special and general theories of relativity proposed by Einstein in 1905 and 1915, respectively. From that time, successive endeavors were performed to grasp and shape the understandings of nature in the overt pursuit of rationalism, that is, by emphasizing predominantly aesthetic simplicity and scientific unity with experience-based but logically deductive methods; yet here, experience could merely "suggest the appropriate mathematical concepts," while remaining "the sole criterion of the physical utility of a mathematical construction" (Einstein 1934, pp. 273–74).

Einstein believed that the real laws of nature could be discovered by seeking those with the simplest mathematical formulation (Norton 2000). Indeed, toward Einstein's grand quest for the ToE, breathtaking leaps were realized in mathematical physics and cosmology via atheistic intuitive research. Such advances included, but were not limited to, the evolving modes of the universe from a singularity of infinite density and gravity (Friedman 1922), the quantum-gravity mechanism of the big-bang origin (Hartle and Hawking 1983; Hartle et al. 2008), and, more strikingly, the cosmic quantum-brain dynamics (Hameroff and Penrose 2003, 2014; Penrose and Hameroff 2011; Penrose et al. 2011). The last discovery was revolutionary. It proved that our universe and mind accommodate not only the various levels of matter in the physical regimes, but also the spiritual evolution of consciousness. Recently, observations showed that our universe is at an expanding stage, homogeneous and isotropic, on sufficiently larger scales of more than about 250 million light-years* (Pettinari 2016, p. 9, 37). In addition, the Planck Mission of the European Space Agency (ESA) provided the most exact age of the observable universe, 13.799 ± 0.021 Ga† (Planck Collaboration 2016).

Notwithstanding the above, it was assumed that the observed universe might only be one causal patch of a much larger unobservable reality. Outside the current cosmological horizon, there might be unknown spheres that cannot be observed. Even for the observed part itself, there were essential observational and theoretical questions left (Carroll 2005; Pettinari 2016, pp. 37–38):

- What is the measurable existence and property of the singularity?
- Is there sufficient observational evidence about cosmic homogeneity and expansion?
- Are the primordial cosmic inflations and the induced density fluctuations even possible?

---

*1 light-year = 9.46×10^{12} km.

†1 Ga = 10^{9} years (i.e., a billion years), or 3.16×10^{16} seconds.

- Is our universe a single and sole *uni*verse, or just one unit of a *multi*verse where our world resides?
- Is the quantum-gravity mechanism determinant for both the big bang and consciousness?
- Is the universe conscious like a gigantic, widely dispersed human brain?

More importantly, and different from the previous theistic base, laws of modern physics are no longer ascribed to God. At this new stage, what is the underlying unified paradigm, without the need of the miracles performed by God, that can comprehensively describe the cosmos-humanity coherence and integrity?

No doubt, modern cosmology owned a philosophical tone in considering the evolving principle of all natural things (Smeenk 2013, pp. 607–52). While the evolution originates from the ontological nothingness (voidness or emptiness) to beingness (existence or fullness), and finally returns to nothingness, the philosophy of cosmology concerns the foundational assumptions and consequences of universal theories and observations (Zinkernagel 2014). Specifically, it offers a spirited defense of the natural connections between ontology and cosmology, named onto-cosmology, the study of beingness. Figure 1.1 (p. 8) provides a sketch for the onto-cosmological study of natural things.

On the one hand, ontology studies the nature or principle of existence, encompassing aspects like qualitativeness, quantitativeness, and relativeness to others (Andina 2014, p. 14). Its aim is to offer answers that elucidate the concepts of what, (who)ness, howness, and whereness. It is concerned:

> with the objects of knowledge, with reality considered in the widest, most profound, and most fundamental aspects, under which the human mind conceives it; with the being and becoming of reality, its possibility and its actuality, its essence and its existence, its unity and plurality; with the aspects of truth, goodness, perfection,

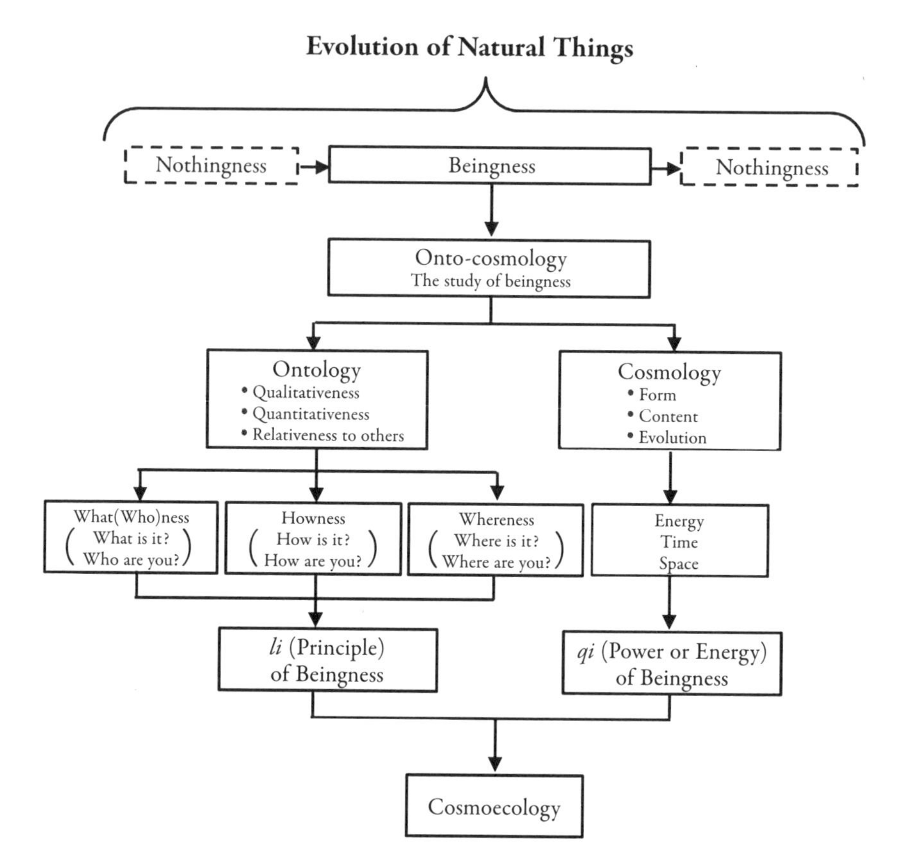

Figure 1.1. Onto-cosmological study of natural things

> beauty, which it assumes in relation with our minds; with the contingency of finite reality and the grounds and implications both of its actual existence and of its intelligibility; with the modes of its concrete existence and behavior, the supreme categories of reality as they are called: substance, individual nature, and personality; quantity, space and time, quality and relation, causality and purpose. (Coffey 1918, p. 23)

On the other hand, cosmology studies the form, content, and evolution of existence by providing answers to elucidate the concepts of energy, time, and space (Smeenk and Ellis 2017). In exploring the universe, cosmology makes use of "the data of direct experience and the conclusions established by the analysis of these data" (Coffey 1918, p. 19). Thus, succinctly speaking, the word *onto-cosmology* refers to the unitary wholeness of both the ontological principle (*li*) and the cosmological vital power or energy (*qi**) of reality. While *li* is a key philosophical concept in neo-Confucianism, *qi* is a foundational concept that supports both neo-Confucianism and Taoism. The two concepts provide influencing ideas about the nature of reality and ethics, shedding light on the relationship between the individual and the cosmos. Their synthesized concept is identified as feng-shui† in Chinese literature, which is comparable to the concept of cosmoecology.

Up to this point, human history from the start of the twentieth century has witnessed two concentrated attempts toward shaping a prototypical paradigm of onto-cosmology. One is atheistic, and the other is theistic or deistic.

ONTO-COSMOLOGICAL PARADIGMS

Atheistic Panquantumism

In modern physics, quantum-gravity formalism was thought to be the cutting edge in dealing with the interrelationship among space, time, matter, as well as consciousness. The constructed models, for example, string/M-theory or loop quantum gravity, were assumed to reconcile two categorically incompatible theories, that is, general relativity for vast-scale space-time and quantum mechanics for infinitesimal space-time (Glattfelder 2019). Such models are mathematically sophisticated but also dubious since they employ either 11-dimensional vibrating

*Chan (1969, p. 634) incorrectly translated *qi* as *li*, a material force. Yet in Chinese, the word *qi* refers to a kind of vital power (or energy) which is a scalar, rather than a force, which is a vector. Also note that the dot product of a "force" and a "displacement" gives rise to "power," defined as the "energy" per unit of time in physics.

†The original meaning of *feng-shui* refers to the harmonic coexistence of all entities with their surrounding environment.

strings or fabric-woven spin foams, both of which are unable to be validated experimentally.

However, philosophically speaking, the scientific advance in modern physics is fundamental for an atheistic onto-cosmology of reality. The quantum-gravity model was mainly formulated by Hawking and James Hartle (Hartle and Hawking 1983; Hartle et al. 2008). The model described a big-bang origin of the known universe. It postulated a hemispherical Euclidian 4-D singularity and stipulated a path-integral over a complex (both real and imaginary) time, where space-time is finite but without any boundaries along the imaginary direction. This was a crucial step in modern science in answering three fundamental questions (Hawking and Hertog 2018):

1. Whatness: What caused the big bang?
2. Howness: How does the expansion keep going eternally without an end?
3. Whereness: Where is God, considering there is no time before the big bang?

This model was contrary to the long-standing historical and prehistorical theistic cosmogony, in which a God-creator was assumed to initiate the universe. Instead, it concluded that the universe originates from a quantum-gravity singularity (Hartle et al. 2008). At the singularity, the cosmic quantum wave function was found to be finite and nonzero (Leplin 1997, pp. 149–50). Thus, space-time was found somehow able to evolve in the forward direction in time, passing through the singularity, leading to a big-bang scenario (Hawking 1996; Penrose 2005, 2012, 2014). Consequently, the creation and evolution of the universe merely follows a physical mechanism, rather than being driven by a theistic prime mover. To be more precise, "science can explain the universe without the need for a creator," and "there is no aspect of reality beyond the reach of the human mind" (Raghav 2016, pp. 71, 77). That is to say, "there would be no singularities at which the laws of science broke down and no edge of space-time at which one would have to appeal to God or some new law to set the boundary conditions for space-time" (Hawking 1988, p. 136). Unambiguously, it was the quantum-gravity mechanism, rather than a God, that was claimed to be the primal source for the universe to expand outward from an initial "quantum-gravity" singularity to a terminal "black-hole" infinity (Hawking and Penrose 1970; Hawking 1996; Penrose 2012).

The expansion of the universe was later elaborated upon by Penrose, who suggested a model of conformal cyclic cosmology (CCC) to answer the three fundamental questions of whatness, howness, and whereness (2005, 2012, 2014) in stating:

1. A theistic prime mover to create the universe is unnecessary.
2. The big bang is triggered by the fluctuations of the nonzero quantum-gravity wave functions at the singularity.
3. The universe experiences cyclic big bangs with every cycle (an eon; a billion years) evolving from an initial "quantum-gravity" singularity to a terminal "black-hole" infinity; the two geometries at both the initial and the terminal states always keep identical in a conformal rescaling invariance, and thus, the universe is able to pass through the singularity and keep the cycles going forward eternally without an end.

Later studies confirmed that the CCC formalism agrees with astronomical observations (Gurzadyan and Penrose 2010), though there had appeared some negative arguments regarding its theoretical analysis of the experimental data, on its lack of the exponentially growing inflation phase after the initiation of the big-bang singularity, and on its prediction of an unrealistic time scale (beyond 10^{100} years) (e.g., Moss et al. 2011).

In addition to the cosmological CCC theory, the quantum-gravity formulation brought about another breakthrough during the turn of the twenty-first century. Penrose and Hameroff connected brain biomolecular processes to the cosmic fine-scale structure upon a transformation, known as the orchestrated objective reduction (Orch-OR), from a big-bang "imaginary time" into a speculative "imaginary space" (Penrose 1989; Hameroff and Penrose 2003, 2014; Penrose and Hameroff 2011; Penrose et al. 2011). Such a transformation adheres to the coherence of the quantum wave functions between the macrocosmic universe and microcosmic brain neurons (Penrose et al. 2011). It is worth mentioning that, at least to some extent, the resultant Orch-OR simulations (e.g., Hameroff and Penrose 2014) explained the experimental signals of the brain EEG waveforms in different states of human consciousness (e.g., Campisi et al. 2012).

Evidently, the macroscopic universe and the microscopic human brain are thus unified by standing on the solid footing of the quantum-gravity principle. This unification between what happens in modern cosmology and neuron biophysics exposes a contemporary ontological panquantumism. The philosophy reiterates the belief of Johannes Kepler in a divine-human universe, on account of the equivalence of God with the quantum-gravity conception at the big-bang singularity (McGrath 2006, p. 214).

Theistic Panentheism

In the West, the theistic integration of matter and spirit was first proposed by Teilhard. He was a French Jesuit missionary in China from 1923 to 1946. There, he was much more renowned as one of the few leading paleoanthropologists in the discovery of the Peking man's skull in 1929.

Building on the Christocentric belief, yet with a commitment to science, Teilhard developed "an original system of thought which may be placed at the frontiers between science, philosophy, theology and mysticism" (Udías 2005). His system rested upon the physical-biological-spiritual pattern of evolution driven by cosmic energy. The energy was assumed to be divided into two distinct components, mind and matter. They spread respectively through two layers, "the within" and "the without," of the world (Teilhard 1975a, pp. 25, 64; 1975b, pp. 134–47).

One component was named the "tangential energy," and the other one was called the "radial energy." While the former links an element with all the others, the "with-out," of the same order, the latter synthesizes the element toward ever greater complexity and centricity through differentiating and ultimately drawing the "with-in" from Alpha to Omega. This final point transcends material evolution and reaches the "God-Omega," also named the "Ω-Point" (Teilhard 1964, p. 116; 1999, p. 186). In the model, Teilhard used an integrated term, the Ahead, to name both Alpha and Omega, which was "existing since the foundation of the cosmos and presenting in the cyclical process of the universe," and behaved as not only "the ground of all existence" but also "the beckoning Cosmic Apex" (Murrell 2022).

On the one hand, the Ω-Point was said to be the endpoint that the universe is moving toward, a point at which the whole of nature has transformed into Jesus, the perfectly synchronized God at all moments of life. On the other hand, the point was defined to constitute "the maximum collective consciousness," named the noösphere, inscribed within the biosphere above the lithosphere, atmosphere, and hydrosphere of the Earth, on which and beyond all things are interconnected (Mesle 1993, p. 106).

According to Teilhard, the "radial energy" exerted a cosmic attractive action, described as "zest" or "love," on the evolutionary process between the future and the present; the process would eventuate "in deeper or greater being" to make "all things to converge toward" the Ω-Point, in a consummation permeating "the whole thing" of evolution; the permeation could be implemented by "radiating back from the future into the present" through humans' resonating "with the universe as a whole as the outer form of our own inner spirit"; and, therefore, the unfolding of nature should surpass the fundamental subjective-objective dualism by depending entirely on how to recognize the universe as a single energy "in an integral way, not as just objective matter but as suffused with psychic or spiritual energy" (Bridle 2016; Udías 2016).

Though incompatible with the Christian orthodox theodicy, such a neoclassical evolutionary process still retained the justification of an omnipotent and omni-benevolent God and thus maintained the transcendent ontological ground of Christian theology. The onto-theological kernel, the so-called Ω-Point, became the logo of the divine power and sovereignty in which a cosmic Christ is the "axis and terminus" of the process, also "the ultimate psychic center in whom all creation will be unified" (Vale 1992; Canale 2005). This process was regarded as a teleological evolution, through a successive governance of an enhancing collective consciousness, over the various levels of all things (Ghislain 2009). The evolution happened in a "divine milieu" of cosmological integrity and interconnectedness (Cobb and Griffin 1976, p. 76; Cavedon 2013), and followed a fundamental cosmo-theological law of "complexification-consciousness," a cosmic movement "complexifying and deepening intelligence or subjectivity . . . further into the depths of consciousness, or interiority" (Sideris 2017, p. 121).

As an alternative to classical theism, Teilhard's philosophy was panentheistic. It integrated both the transcendental theology (i.e., onto-theology and/or cosmo-theology), as proposed by Immanuel Kant, and the wide-ranging scientific and spiritual findings of the cosmological evolution. Different from the classical theism which asserts a traditional ontological argument that ALL *is from* God, that is, the world is an "extrinsic" property of God (e.g., Göcke 2013), Teilhard stressed that ALL *is in* God, that is, the world is an "intrinsic" property of God (Cooper 2006, p. 26). Undoubtedly, this was a different thesis from the kind of Catholic pantheism which claims that ALL *is* God, that is, that the world is ontologically identical to God (Culp 2021).

In spite of its inconsistency with Catholicism, Teilhard's scheme of evolution was broadened and deepened by some farsighted priests, including Berry. He followed Teilhard's forward-looking mind-matter synthesis and proposed an onto-cosmological doctrine. It featured a psychic-physical (or mind-matter) integral, briefly, a "functional cosmology." The kernel of the integral lies in the cosmogenesis, which was suggested to be "characterized by the increasing complexity, consciousness, and cephalization, i.e., the development of the central nervous system toward large-brained humans," while Teilhard's synthesis was recast by a noöspheric transformation which could "powerfully evoke positive human change and an ecological praxis" to arrive at a cosmic "ecozoic" stage, that is, "a dawning geological era and higher stage of cosmic evolution marked by a new mode of human involvement in the Earth system" (Sideris 2017, pp. 119–23). Berry believed that, during the transformation, the physical and psychic dimensions of the cosmic energy are folding into each other, yet neither is collapsible into, nor is separable from, the other (Mickey 2016, p. 41).

THE UNIVERSE: AN ANIMATE BEING?

Both scientific panquantumism and theistic panentheism point in the same direction: Nature is characterized not only by a physical regime and evolution, but also by a spiritual essence and perfection. That is to say, our universe may indeed be a living organism like a human creature. Table 1.1 details the basic properties of the observed universe described by the two paradigms and their comparisons to an animate human being.

TABLE 1.1. COMPARISON BETWEEN THE OBSERVED UNIVERSE AND THE HUMAN BEING

Organism	Observed Universe		Human Being
	Panquantumism	Panentheism	
Component	existence/consciousness	matter/mind	body/consciousness
Evolution	conformal cyclic	cyclical	biological cyclic
Environment	dark matter/energy	divine milieu	sunlight-air-water
Initiation	quantum-gravity singularity	alpha point	fertilized egg
Growth	all-directional, synchronous		
Transformation	rescaling invariance	collective consciousness	mental ability
Termination	black-hole infinity	omega point	aging to death

In view of either of the two paradigms, our universe follows an uncertainty-determined causal determinism, simply, the cause-and-effect law upon random probabilities of initiation, which says:

1. A singularity in vacuum gives a big-bang explosion at some unpredictable quantum-gravity state.
2. The explosion expands space-time in all directions synchronously to form the observed universe at the current stage. And,
3. The expansion will eventually stop and the universe will contract and collapse to a future black-hole infinity (the heat-death state of the universe) which conforms and rescales the invariable geometrics back to the singularity.

Amazingly, this cosmological process reproduces the progresses of a human creature growing from a fertilized egg initially, which comes into being from emptiness, develops through the phases of gestation in the uterus, and grows after birth but decelerates at the end of puberty to the last stages of aging and death before returning to dust. Perhaps such astonishing similarities between the macrocosmic universe and the microcosmic human body imply a holographic* unity of Heaven, Earth, and Humanity, giving rise to an unusual theory of everything (ToE).

This book aims at describing such a unity based upon developing Swimme's collective consciousness of modern cosmology. This approach consists of two intertwined theories. One theory is the Berry-Swimme trinity of ontological cosmogenesis. It includes differentiation (or diversification), autopoiesis (or self-making), and communion (or interconnectivity) (Mickey et al. 2017, p. 150; Swimme and Berry 1992, pp. 71–73). The other theory is the explication of cosmic creativity through Swimme's evolutionism that features a set of twelve cosmogenetic powers (Amberg 2011, pp. 7–39; Le Grice 2011, pp. 222–23;

*The term *holographic* means wholeness is present within any components of an entity; by contrast, *holofractal* means wholeness is self-similar at any scale of an entity.

Swimme 2017a, b; Swimme and Anderson 2004). To reach that goal, the following methodologies will be employed:

- First, empirical approach. Rather than theoretical research that brings challenges in terms of applicability, this first approach relies on primary sources (e.g., original Chinese classics) and secondary ones (e.g., publications available in the library or online).
- Second, inductive approach. Different from deductive research that emphasizes theory testing via datasets, surveys, or quantitative analysis, this second approach synthesizes Eastern philosophical wisdom and Western scientific achievements.
- Third, scientific approach. Instead of metaphysical research over and beyond physics with speculative consciousness, this third approach is based on the integration of scientific achievements, such as Leibniz's binary numerals borrowed from the I Ching's hexagrammatic categories (e.g., Legge 1963; Sung 1973), modern cosmological theories (e.g., Friedman 1922; Belenkly 2012), the Hawking-Penrose theorem of quantum-gravity cosmology (Hawking and Penrose 1970; Hawking 1996), and Penrose's CCC model (e.g., Penrose 2012), as well as the Penrose-Hameroff Orch-OR formalism of brain consciousness (Hameroff and Penrose 2003, 2014).

In our pursuit of an answer to Einstein's quest for a ToE, our journey will review the ontological and cosmological paradigms in history, and then turn toward a more profound comprehension of the subject. In chapter 2, we pivot our focus to establish a historical foundation by delving into the background that has shaped comprehensive theories and thoughts in explaining the intricacies of the universe. Such a strategic shift will ensure that a solid groundwork has been laid, setting the stage for the exploration that will unfold in the subsequent chapters.

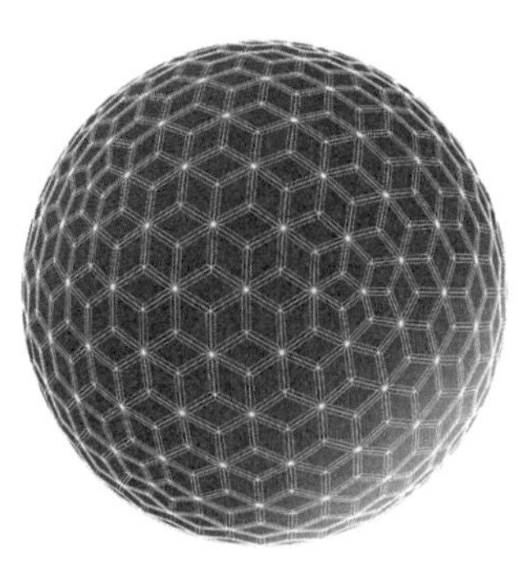

2 Historical Background

Homo sapiens, the earliest modern humans, came into being in Africa about 200,000 years ago. They continuously migrated worldwide through Eurasia (Stringer and Andrews 1988). Historically, there emerged four cradles of ancient civilizations on the planet (e.g., Oliphant 1993):

- Middle East
  - **Mesopotamia:** 5000–2900 BCE (Kreis 2014)

- North Africa
  - **Egypt:** 3100–332 BCE (Kreis 2006)

- Europe
  - **Minoan (Mediterranean Sea):** 2000–1450 BCE (Dorling Staff 1994, pp. 24–29)
  - **Mycenaean (southern Greece):** 1600–1200 BCE (Kreis 2009)

- Asia
  - **Indus Valley (India):** 3300–1300 BCE (Possehl 2002, pp. 23–88)
  - **Yellow and Yangzi River valleys (China):** 7000 BCE–1840 CE (Schirokauer and Brown 2011, pp. 2–4)

Of these four, Chinese civilization had survived relatively intact from the prehistorical era down to the nineteenth and twentieth centuries (e.g., Gernet 1996). By contrast, all the other ones had developed either in tandem owing to cultural diffusion (e.g., between Egypt and the Indus Valley) or with traits inherited or passed from one to the other (e.g., from Minoan to Mycenaean). Following the fall of the Mycenaean civilization in ancient Greece, Greek culture emerged in the eighth century BCE at the end of the Greek Dark Ages (1200–800 BCE). The emergence was associated with the first Olympic Games that were held in Athens in 776 BCE (Hall 2007, pp. 41–66). By the end of the fourth century BCE, the language and values of Greek civilization spread across Europe. Although the stronger Roman Empire defeated Greece before the first century BCE, Greek achievements in such things as education, art, literature, philosophy, and science had pervaded the Roman Empire and taken root in Europe (Needham 1998, p. 28). Western civilization was thus formed and came to thrive on the continent (Mazlish 2004, p. 3).

During the long history of world civilizations, there are four stages in onto-cosmological philosophy: ancient onto-cosmology, metaphysical onto-cosmology, mechanistic cosmology, and physical onto-cosmology.

ANCIENT ONTO-COSMOLOGY

Ancient onto-cosmology refers to the onto-cosmological concepts of humans before the eighth century BCE. At that time, the universe was mostly understood through notions of magic (Schombert 2016a). Supernatural themes, creation myths, gods, and divine or semidivine figures such as Gaia, the ancestral mother of all life and the primal Earth goddess of the pre-Greek origin story (e.g., Beekes 2009, pp. 269–70), and Brahma, the Indian god of creation (e.g., Cartwright 2015), as well as Pangu and Nvwa, the mythologically primordial Chinese figures (e.g., Walls and Walls 1984, p. 135), were assumed to drive, or be responsible for, natural phenomena.

Nevertheless, there was an exception in Chinese history. From about 4,800 years ago to the ninth century BCE, there lived two great characters in the central plain of China, Fuxi and King Wen of Zhou (simplified as King Wen in the following text). They were acknowledged for pioneering the creation of the I Ching, the oldest and most intricate system of numerological divination in the world, used to elucidate ancient atheistic ontology and cosmology (Antinoff 2010, p. 94). The essence of this masterpiece includes the following ideas:*

- *Qi* divides into *yin* (called *yinqi*) and *yang* (called *yangqi*) which are opposite in polarity, while mutually integrated, interdependent, consumed, and transformed (Zhou 1990, pp. 3, 26).
- The primordial *qi* can transform or distribute into subordinate components of *qi* to drive *yi* (change, becoming, or evolution) of all entities.
- *Yi* divides into five types: congruent becoming, free self-becoming, forced self-becoming, interactive becoming, and harmonic becoming (Ma 2016).†
- All *yi* brings about an ecological harmonization of reality, the oneness of heaven and humanity.

According to the I Ching, heaven is metaphorized as the macrocosmic entity of humanity, while humanity is metaphorized as the microcosmic entity of heaven, with the process of mutual sustenance entirely demonstrated by both the symbols of the posterior-heaven eight trigrams and the theory of five phases (Dong, 206 BCE, chapter 42). No

---

*In this book, we have chosen to italicize the six crucial words for which there are no accurate English expressions, because, if named, "the names are unable to convey the true nature or essence of the things to be named," as the Tao Te Ching says. Therefore you will always see the words *yin*, *yang*, *li*, *qi*, *yi*, and *dao* in italic.

†The first three types are from (*a*) Zheng (n.d., p. 1); (*b*) Cheng (2006); and (*c*) Wilhelm (1990, pp. 280–85), while the last two are from Cheng (2006). It deserves mention that Wilhelm misinterpreted the first three types.

later than the Warring States period (475–221 BCE), this philosophy gave rise to the rules of cosmo-ecology (Shao 2004, p. 77). It deserves mentioning that, until the earlier time of the early Zhou dynasty (1046–256 BCE), the theory and the rules laid a solid foundation for one of two ancient medical systems, traditional Chinese medicine. Prior to it, the other system, the Mesopotamian tradition, was already well established (e.g., Teall 2014).*

METAPHYSICAL ONTO-COSMOLOGY

Metaphysical onto-cosmology includes a couple of dominant philosophical systems in the era from the eighth century BCE to the sixteenth century CE. One is the metaphysical onto-theological cosmology that was formed in the West. The other is the metaphysical onto-numerological cosmology that arose in the East.

Onto-theological Cosmology

In the West, Greek Pythagoreanism developed from Babylonian mathematical techniques and astronomical records. Pythagorean ideas exerted substantial influence on Plato, Aristotle, and those who followed them.

Plato developed the ancient Greek cosmology by suggesting an onto-theological concept. He proposed that the cosmos was created by an ontologically conscious creator who is "always unchangeably real" (Cornford 1997, p. 39). Planets were designed to cosmologically rotate in their perfect circular orbits around the Earth (Cornford 1997, p. 57). Aristotle supported Plato's ontological understanding of reality by insisting that an unmoved Being drives the universe unconsciously (Hakak 2002). At the same time, Heraclides was also in favor of Plato's

*However, some specialists in the history of Chinese medicine and Chinese philosophy may disagree with this statement because, according to their view, the time of "until the early Zhou dynasty" cannot be proven (J. F. M. des Jardins, personal communication, May 2019).

cosmological model. He proposed a complete geocentric description of the universe. The model argued that all the celestial bodies, stars, planets, the Sun, and the Moon were orbiting the Earth. Based on this geocentric model, Aristarchus set up the first heliocentric model (Schombert 2016b).

Scientific knowledge emerged from the scientific reasoning and investigation done by natural philosophers in ancient times. Plato's onto-theological cosmology climaxed during the period from the second century BCE to the second century CE. Hipparchus systematically studied Babylonian astronomical knowledge and techniques (Toomer 1988). He invented trigonometry, discovered the precession of the equinoxes, and divided the circle into 360 degrees of 60 arc minutes (Linton 2004, p. 52; Toomer 1996, p. 81; McCluskey 2000, p. 22). In the second century CE, Ptolemy replaced geocentric cosmology in his *Almagest* with a newer understanding of a circular structure in different cosmic scales (Pedersen 2011, pp. 26–46). The Ptolemaic theory endured until the sixteenth century when Copernicus adapted the onto-theological geocentric model to meet the requirements of Aristarchus's heliocentric universe (Copernicus 1995, trans. Wallis).

Eventually, a so-called transcendental theology was formulated. It included two competing types: onto-theology and theological cosmology (Thomson 2005, p. 7). According to Kant,

> Transcendental theology aims either at inferring the existence of a Supreme Being from a general experience, without any closer reference to the world to which this experience belongs, and in this case it is called cosmo-theology; or at the endeavors to cognize the existence of such a being, through mere conceptions, without the aid of experience, and is then termed ontotheology. (Kant 1872, p. 388)

That is to say, while onto-theology endeavors to "cognize the existence of such a being, through mere conceptions, without the aid of experience," theological cosmology introduces a "new" theistic cosmology which held a "cosmo-theological" belief in view of "a general experience," such as Galileo's groundbreaking telescopic observations made in the seventeenth century.

Onto-numerological Cosmology

Until the end of the sixteenth century, the onto-cosmological philosophy of the I Ching had continued to develop in China. This development was based upon the integration of the philosophies of Buddhism, Daoism, and Confucianism. It reached the culmination around the eleventh and twelfth centuries labeled by the emergence of a definitive metaphysical doctrine, the oneness of heaven and humanity, as mentioned previously.

Buddhist philosophy arose from the teachings of Shakyamuni Buddha and Nāgārjuna Bodhisattva. It recognized *sunyata* (pure emptiness) as the ontological ultimate of reality (Damdul 2019, pp. 345–49). The quest for the cosmological basis of evolution led to the discovery of *pratītyasamutpāda* (interdependent origination) in a series of twelve links (Buswell and Lopez 2014, p. 854), i.e., ignorance, fabrication, rebirth consciousness, name-form, six-entrances, contact, sense, craving, attachment, becoming, birth, and aging-deceasing. This causation process was claimed to be driven by the first karmic element, ignorance (Gethin 1998, pp. 112–32; Sadakata 2009, pp. 19–40, 93–112).

Daoist onto-numerology emerged around the same time. The Tao Te Ching (or Daodejing, known as the Book of Dao and Virtue) was the first masterpiece that became central to Daoism. It was ascribed to a deified semi-legendary philosopher, Laozi (or Lao-tzu), who was believed to have studied the I Ching's fifteen hexagrams (table 3.3, p. 51). His speculative consciousness of the hexagrammatic reality was treated as the guide to the establishment of Daoism (Yang n.d.). Concisely, the Tao Te Ching describes an evolving universe in continuous and endless cycles. It argues that the world "revolves eternally without exhaustion, relying on nothing, with the end of one cycle to be the beginning of the next one" (Laozi, chapter 25).

During the Warring States period, the second Daoist masterpiece, the Yellow Emperor's Classic of the Secret Talisman, became popular. It offered an ontological philosophy of cosmo-ecology (Anonymous 700–750 CE; Legge 1891, pp. 257–64; Li 1983). Its essence lies in the phrase "triadic interdependence of three treasures," where the "treasures" refer to Heaven, Earth, and Humanity; and, the "interdependence" happens by following a cosmo-ecological principle (e.g., Zhang and Li 2001, pp. 113–24).* The phrase discloses a dynamically harmonized generation and an overcoming of the processes of both *yin* and *yang*, the two opposite but complementary types of *qi* that drive the regulation, the functioning, and the modes of the energies in the evolution of all entities. Due to these two types of *qi*, the theory of the five phases (that is, wood, fire, earth, metal, water) extends to deal with ten ingredients of *qi* (Q1: *yang* wood; Q2: *yin* wood; Q3: *yang* fire; Q4: *yin* fire; Q5: *yang* earth; Q6: *yin* earth; Q7: *yang* metal; Q8: *yin* metal; Q9: *yang* water; Q10: *yin* water) (Dong, chapter 42).

Kongzi (or Confucius) studied the I Ching and elaborated his thinking about the nature of heaven, of humanity, and of human society, as well as the connections among them. He developed a philosophical conservatism with regard to society and ethics. He completed a collection of ten commentaries on the I Ching, called the Ten Wings (e.g., Rutt 2002, pp. 363–56). They were compiled in the Five Classics (e.g., Nylan 2001, pp. 229–52). His doctrine demonstrated the humanistic legitimacy of duty, obedience, and hierarchy in upholding social ethics, legal practice, and family morality.

Around 1000 CE and no later than the twelfth century, a hybrid philosophy, neo-Confucianism, flourished by transforming traditional

---

*This is the translation of the Chinese phrase *sān cái xiāng dào*. Note that Zhang and Li (2001) translated it wrongly into "mutual stealing among the three powers" for two reasons: first, the word *cái* means "treasure" or "valuable objects," rather than "power" or "energy"; and second, the word *dào* refers to "interdependence" or "mutually making use of" in classical Chinese, rather than its literal meaning of "stealing" or "robbing" in modern Chinese.

Confucianism through a dialogue with Daoism and Buddhism. It harmonized and integrated the Daoist ontological concept of heaven, the cosmological and spiritual foundations of Buddhism, and the humanist ethics and morality of Confucianism (Feng 2004, p. 232; Fung 1948, p. 268). Realistic rationalism, the oneness of heaven and humanity, was reformulated by going beyond Daoist transrational mysticism and radically spurning Confucian ossified thinking (c.f. Dusek 1999, p. 67). This transformed philosophy discussed the evolution of heaven, humanity, and related human society as a whole by tracking a holographic and holofractal process applying to different scales of the universe. Historically, neo-Confucianism had different schools named by alternative terminologies. Two well-known ones were the learning of *dao* (the way) and the learning of *li* (the principle).

After the eighth century, the metaphysical consciousness of onto-numerological cosmology achieved its maturity owing to the contributions of the six most prominent neo-Confucian philosophers: Zhu Xi and the Northern-Song Five Masters: Shao Yong, Zhou Dunyi, Zhang Zai, Cheng Hao, and Cheng Yi. The philosophy they articulated was the metaphysical dualism of *li-qi*, that is, the interplay between *li*, the ontologically unchanging, transcendent principle, and *qi*, the cosmologically dynamic, transformative power of nature. While *li* represents the neo-Confucian *dao* (equivalently, *wuji*), the ultimate reality or the underlying ultimateless that is formless and beyond conceptualization, *qi* indicates the dynamic source or basis to propel *yi*, the change, becoming, or evolution of natural things and the whole universe (Chen 2011; Zheng 2007; Zhu Xi Song Dynasty). Both the principle and the basis were believed to elucidate how all things and the universe arise by following *dao* in order to live a harmonious and balanced life by both embracing the natural order of interconnectedness and aligning with the spontaneous and natural unfolding of existence (Hinton 2013, p. 133). What is more, the paradigm thus constructed assumed the ultimate universal heaven-humanity perfection per the I Ching's mystical hexagrammatic *qi* in its regularity and transformation (Cheng 2006).

A couple of achievements resulted owing to this ancient metaphysical onto-cosmology. One achievement appeared in the first century BCE. It was the lunisolar calendrical system, which employs twenty-four nodal-medial *qi*, known as the solar terms. This set of twenty-four solar terms became a UNESCO Intangible Cultural Heritage in 2016. The terms correlate to the four seasons. The names and the related initialisms are as follows:

- **Season 1, Spring:** J1: Spring Beginning (SpB); J2: Rain Water (RW); J3: Insect Waking (IW); J4: Spring Equinox (SE); J5: Pure Brightness (PB); J6: Grain Rain (GR)
- **Season 2, Summer:** J7: Summer Beginning (SuB); J8: Less Fullness (LF); J9: Grain in Ear (GE); J10: Summer Solstice (SS); J11: Less Heat (LH); J12: Great Heat (GH)
- **Season 3, Autumn:** J13: Autumn Beginning (AB); J14: Heat End (HE); J15: White Dew (WD); J16: Autumn Equinox (AE); J17: Cold Dew (CD); J18: Frost Descending (FD)
- **Season 4, Winter:** J19: Winter Beginning (WB); J20: Less Snow (LS); J21: Great Snow (GS); J22: Winter Solstice (WS); J23: Less Cold (LC); J24: Great Cold (GC) (Liu 2013; Nielsen 2003, pp. 75–76)

The other achievement happened during the eleventh century. It was Shao Yong's treatise, the Supreme World-Ordering Principles, that interpreted the cyclically evolving property of reality. The treatise relied on the taxonomic system of the I Ching's sixty-four hexagrams to present the principles as numeral categories. On the one hand, the ontological pattern obeyed an elegant progression in terms of numerical orderings (Huang 1999, p. 52; Vikoulov 2017). On the other hand, the cosmological basis relied on the *xiao-xi* arrangement of *yao*, the hexagrammatic lines, stacked in the trend of *xiao* (that is, "waning," either yin-ascending or yang-descending) and *xi* (that is, "waxing," either yang-ascending or yin-descending). The distribution is characterized by the alternating categories of 12 and 30, along with a set of four periodicities:

- **Periodicity 1:** *yuan* (Cycle), a period of 129,600 years
- **Periodicity 2:** *hui* (Epoch), a period of 10,800 years, 1/12 of yuan
- **Periodicity 3:** *yun* (Revolution), a period of 360 years, or, 1/30 of hui
- **Periodicity 4:** *shi* (Generation), a period of 30 years, or, 1/12 of yun, where Year is 1/30 of shi, named as *nian*, a period of 360 *ri* (6 *ri* = 6.0875 days), 1 calendar year

Thus, a full Cycle includes 12 Epochs, 360 Revolutions, 4,320 Generations, and 129,600 Years. It agrees with the ice-age cycle of approximately 100,000 years when colder global temperatures led to recurring glacial expansion across the Earth's surface. Within such an ice age, life on Earth experiences a series of four phases, namely, birth, growth, preservation, and decline. Similar phases were proven to have existed in previous ice ages, and also predicted to happen in future ones (e.g., Abe-Ouchi et al. 2013; Zurich 2013).

MECHANISTIC ONTO-COSMOLOGY

The mechanistic onto-cosmology dominated the world during the sixteenth to nineteenth centuries. In this period, the industrial revolution began in Europe with the invention of the mechanical knit-stocking frame at Woodborough in 1589 (Landow 2012). There were two surges of revolutionary changes. The first one started in the United Kingdom and subsequently spread throughout Europe, North America, and eventually influenced most of the world until around 1850. The second one began in the middle of the nineteenth century and resulted in global industrialization.

This industrialization period begat a new theological era, called the Planetary Era (Morin and Kern 1999, p. 5ff). For a few hundred years, European science and philosophy developed with a theological belief in deism: a supreme God behaved as the watchmaker who constructed the universe and then let it proceed on its own. A classical mechanistic cosmology entirely replaced the previous Aristotelian cosmology owing to the unprecedented progress under the umbrella of Christian universalism (Whittmore 1840).

Pre-Newtonian Era

Prior to Newton, Thomas Digges (1999, 2001) extended the Copernican onto-theological heliocentrism to a static cosmological model. He proposed that the universe is

- stationary (i.e., neither expanding nor contracting);
- flat (i.e., no spatial curvatures); and
- infinite spatially and temporally.

Later, Kepler reached a breakthrough by proposing the existence of a divine-human universe (McGrath 2006, p. 214). He believed that the planetary motion in such a reality follows three laws valid in a heliocentric system:

- **First law:** planets move in elliptical orbits with the Sun at one focus.
- **Second law:** a line from the Sun to a planet sweeps out equal areas in equal times.
- **Third law:** the square of the period of any planet in years is proportional to the cube of the semi-major axis of the elliptical orbit.

The first two laws were put forward in 1609, and the third one in 1619.

In the same period, Galileo reexamined and developed Digges's onto-theological model with the aid of celestial observations. He followed the Dutch design of the first telescope in 1609 to create his own, and his fertile and efficient observation of celestial objects opened the door to the present-day observational astronomy (Angelo 2006, pp.

253–54; Frautschi et al. 2007, pp. 3–4). With his keen insight into the character of natural phenomena, Galileo (1610) advocated that:

- the Milky Way is composed of numerous individual stars;
- the Sun has a similar nature in light emission to other stars;
- the surface of both the Sun and the Moon are not smooth;
- Jupiter has four satellites, and Venus has Moon-like phases; and
- all the planets obey the law of inertia: once it starts to run, the motion becomes endless, without the need to maintain any external forces.

Though supporting the Copernican heliocentric theory, these results did argue that the Sun is merely one of numerous stars in the universe. This recognition was against the prevailing heliocentrism at that time. Yet Copernicus's heliocentric model exemplified the Copernican Principle, providing a specific cosmological representation that supports the idea that neither the Sun nor the Earth occupies a central position or holds privileged positions in the universe (Bondi 1952, p. 13).

More significantly, Galileo realized that mathematics should be able to express the great book of nature. Therefore, the *qualitative* descriptions of nature exposed in Aristotle's writings must give way to *quantitative* expressions of formulae coming from precise measurements (Frautschi et al. 2007, pp. 9–12). For this reason, he applied mathematics in physics with the "Galilean equations" (MacDougal 2012, p. 33). For example, he demonstrated that the most critical relationship in classical mechanics is among parameters, like height, speed, acceleration, and time, all required for objects to fall to the ground.

Newtonian Era

Newton made the most significant achievements in mechanistic cosmology. He explored the connections between specific forces and resultant movements and developed the kinematics of planetary motions. In addition to his designs for a reflecting telescope, he not only applied calculus to reformulate Kepler's three laws but also systematized three laws of motion and the law of universal gravitation. The celestial mechanics thus developed expounded exactly the interactions among force, mass, speed, and distance (Kelvin and Tait 1912, pp. 240–48; Newton and Machin [1729 in Latin, pp. 19–21], 1846). The four laws are as follows:

- **The first law of motion** (principle of inertia): the velocity of an object is always constant if the net force, the vector sum of all the forces acting on the object, is zero.
- **The second law of motion** (principle of force and motion): in an inertial reference frame, the rate of change of the momentum of a nonzero-mass object is directly proportional to the force applied to the object, and this change in momentum takes place in the direction of the applied force to produce an acceleration.
- **The third law of motion** (principle of action and reaction): there is always an opposed equal reaction to every action, while the force of the action is always equal in magnitude and opposite in direction to that of the reaction.
- **The fourth law of motion** (principle of universal gravitation): every mass attracts every other mass, and the force of attraction between two masses is directly proportional to the product of the masses and inversely proportional to the square of the distance between the centers of the masses.

Newton used these laws to test and generalize Kepler's laws by solving a two-body problem. Unprecedented results included that ellipses are not the only orbital paths; instead, orbits can be either bound (ellipses and circles) or unbound (parabola and hyperbola); in addition, the mass of a central body can be readily calculated if surrounded by a small body with a known orbital period and a known orbital distance (Frautschi et al. 2007, pp. 457–73).

Based on the observed celestial orbits of planets and comets, Newtonian mechanics made it possible to develop an onto-theological and mechanistic cosmology. Rooted in the mathematical formulation of the Copernican principle, the paradigm conceptualized the following:

> the light of the fixed stars is of the same nature with the light of the sun, and from every system light passes into all the other systems: and lest the systems of the fixed stars should, by their gravity, fall on each other mutually, he hath placed those systems at immense distances one from another. (Newton and Machin 1729, pp. 387–93)

This notion enhanced Digges's views on space-time leading to an isotropic and homogeneous model by taking it for granted that

- an impersonal God is ontologically the first cause of reality; as a result, God creates space and time; and
- nothing is unique in the universe (i.e., neither planets nor stars nor galaxies, occupy any particular or privileged positions).

That is to say, the cosmic mass distribution in the universe is uniform (i.e., spatially homogeneous, and isotropic) for a family of typical observers who are co-moving with the surrounding matter (Frautschi et al. 2007, pp. 457–73; Newton and Machin 1729, pp. 387–93).

Post-Newtonian Era

After Newton, the onto-theological mechanistic cosmology was gradually weakened and gave way to other scientific accounts. For example, in place of Newton's theistic origin of the solar system, Kant posited that not merely this system, even the whole Milky Way is just one of many galaxies formed from vast nebulae (Kant 2008, pp. 36–37). As one of the earliest cosmologists, Kant applied Newtonian mechanics and mathematics to prove the existence of the gravitational collapse and "invisible bodies," that is to say, black holes (Laplace 1799). He also translated the geometry-based Newtonian mechanics to calculus-based "celestial mechanics" (Laplace [1829], 1969).

Afterward, Pierre-Simon Laplace continued the studies and provided the first scientific description of how either the solar system or the universe was formed. The model suggested that cosmic clouds are driven by gravitational forces to initiate angular momenta; the angular motions act on different clouds to form disklike planetary distributions around the Sun; the same mechanism is applicable to the formation of galaxies, galaxy groups, or clusters, and other celestial realities (Laplace 1830).

Until the nineteenth century, scientific research developed in an unprecedented way, far beyond the regimes of astronomy and cosmology. Such fields included, but were not limited to, analytical mechanics, fluid dynamics and aerodynamics, optics, chemistry, thermodynamics, electricity, and magnetism (Serway and Jewett 2014; Williams 2018).

Also of note in our story is the fact that the Opium Wars broke out in the 1840s. Equipped with a powerful military force, the West opened the door to China. The imperialist invasion made it possible for Western missionaries to enter China freely. The evangelization resulted in a surge of Sino-Western contacts. Ultimately, Christianization failed as a result of the unshakable cohesion of Confucian ideology among Chinese people. Notwithstanding this fact, the dissemination of Western science and culture has never been suspended in China for two reasons:

- an internal cause: the enthusiasm of the Chinese people to catch up with the West in all fields of knowledge (e.g., economy, politics, science, education, particularly military force); and
- an external cause: the missionaries' drive to push forward all kinds of revolutions in China to make it a Westernized country in the campaign of Christian universalism.

Some examples of the scientific and cultural dissemination include astronomy and mathematics via Matteo Ricci, education and publication of Christian texts via Robert Morrison, political reform and higher education via Timothy Richard, and diplomacy and university organization via John Leighton Stuart (Ashton 1948, pp. 58–93).

PHYSICAL ONTO-COSMOLOGY

Physical onto-cosmology was a philosophical ideology characterized by a large-scale rejection of theism, at least among the scientists in academic disciplines in astronomy, physics, mathematics, biology, archaeology, and genetics. It came into being during the 1900s when modern physics began to push forward the frontiers of an atheistic philosophy regarding the reality of nature. This modern philosophy was reflected by a set of three fundamental theories identified nowadays as the cornerstones of modern physics. Without pointing to any divine traces, the three theories included:

- quantum theory, proposed by Planck in 1900;
- special theory of relativity, suggested by Einstein in 1905;
- general theory of relativity, also suggested by Einstein in 1915.

Einstein made three breakthroughs in the formation of modern physics (Kragh 2015, pp. 441–55):

- predicting the photoelectric effect by applying Planck's quantum theory;
- extending and amending low-speed Newtonian laws to relativistic cases by making use of his special theory of relativity, while abandoning the ether hypothesis and affirming the mass-energy relationship; and
- explaining the precession of Mercury's orbit from his general theory of relativity.

Soon after Einstein's work, Alexander Friedman investigated the relativistic theories in curved space-time (which can be seen in figure 2.1) proposed by Einstein's collaborator, Willem de Sitter. Friedman identified three possible paths of a non-static universe in its cosmological evolutions (Belenkly 2012):

- expansion at decelerating rates from a zero-radius singularity until an inflection point after which the expansion accelerates;
- expansion at accelerating rates from a nonzero initial radius, or expansion in an oscillating style; and
- expansion from zero radii and back to zero radii in a periodic scenario.

Later, Georges Lemaître applied Friedman's model to studying an expanding universe (Lemaître [1927 in French], 1931). He hypothesized that the universe should originate from a primeval atom or a cosmic egg, that is, the big-bang assumption (Midbon 2000). The hypothesis was observationally verified first of all by Edwin Hubble (Hubble 1929) and then by a broad range of astronomical data (Wright 2019). Although

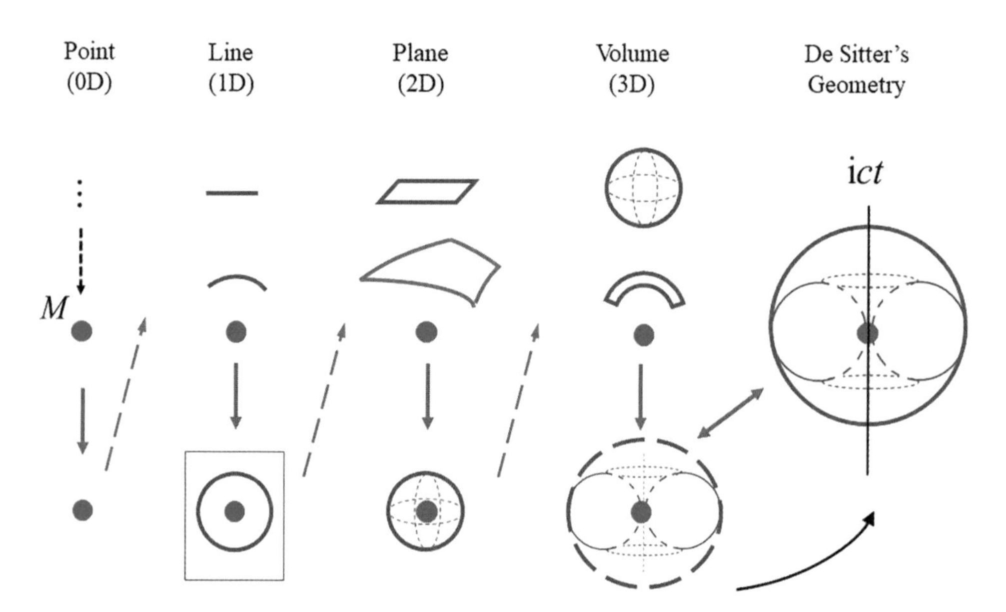

Figure 2.1. Visualization of absolute space and De Sitter's curved space-time

there existed alternative atheistic cosmological models, such as plasma cosmology (e.g., Peratt 1995), the big-bang model became the prevailing standard one. As described in Hawking (1996), the model implied that

- the primal source of a big-bang universe depends on quantum fluctuations;
- the quantum process drives the universe to grow and expand outward from a singularity; and
- a God is unnecessary from the beginning of the universe.

Consequently, considering the historical backdrop of onto-cosmology, it becomes evident that the ontological primal source undergoes a transformation, shifting from theistic to atheistic perspectives in tandem with human knowledge progressing from prehistoric insights to scientific advancements. The prevailing consensus acknowledges that the necessity of a deity diminishes, starting from the inception of the universe. Notably, the Hawking-Penrose quantum fluctuations have been recognized as the cosmological driving force behind the universe's evolution, originating from a singular point.

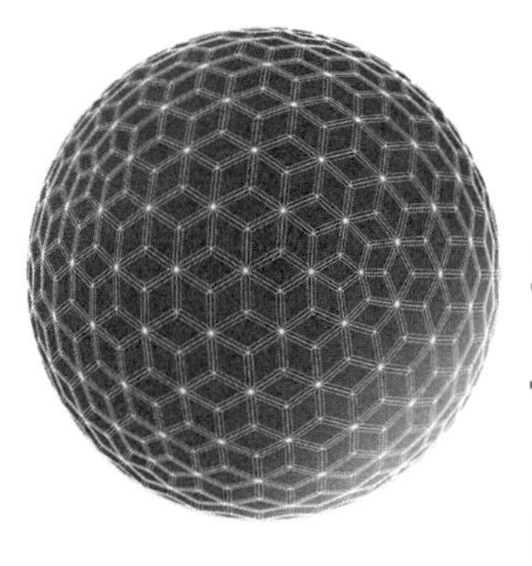

3 The Wisdom of Eastern Philosophy

The wisdom of Eastern philosophy is clearly manifested in Daoist ontology and neo-Confucian cosmology with regard to the oneness of heaven and humanity. As articulated extensively and intensively in the I Ching, the wisdom expresses the cosmic correlation between heaven and humanity: "Things of the same kind summon each other, those with the same qi coincide together, and those with matching sounds resonate" (Chen 2002, p. 1369; my translation).

The name of "I Ching" contains two Chinese characters: "I" (that is *yi* in the Wade-Giles system of romanization) and "Ching" (that is, *jing* in the same system). The first character refers to an ontological "change," "becoming," or "evolution" at either macrocosmic or microcosmic scales, while the second word means "scripture" or "book." According to the I Ching, *yi* expresses the consecutive sequences of changes for all the entities: "(the process of) production and reproduction is what is called *yi*" (Sung 1973, p. 281). As mentioned in the section on "Ancient Onto-cosmology" in chapter 2 (p. 19), there are five types of *yi* driven by the primordial *qi*: congruent, free self, forced self, interactive, and harmonic. All these five types of *yi* are not only characterized by "thoughtlessness, effortlessness, quietness, and stationariness" (Sung 1973, p. 295), but are also able to be "induced to an activity producing responses that penetrate forthwith to all the phenomena and events of the universe upon stimuli" (Sung 1973, p. 295).

It deserves mention here that another term, *dao* (or *tao*), in the Ten Wings of the I Ching has a definition different from Daoism and neo-Confucianism. It is defined as a designation for the unity of *yin* and *yang*: "the appellation of the unity of one *yin* and one *yang* is called the *dao*" (Sung 1973, p. 280; Zhu Xi 1992, p. 141). That is, in the lexical domain of philosophy, this word, *dao*, has a dialectical interpretation with both *neihan* (an ontological connotation) and *waiyan* (cosmological denotation). Thus, it presents a dialectical monism (or, dualistic monism) by holding such an onto-cosmological position that reality expresses itself in dualistic terms but ultimately a unified whole.

DAOIST ONTO-COSMOLOGICAL PHILOSOPHY

Classical commentaries on Daoist philosophy are scattered in books, pamphlets, and papers. Most of them are in traditional Chinese characters. Present-day Chinese scholars have made a few authoritative reviews or compilations of Daoist metaphysical onto-cosmology. For example, Zhang (1982, pp. 17–22) discussed the ontological connotation of *dao* in his *Outline of Chinese Philosophy*. Ren (1998, pp. 398–402, 253–61) elaborated Zhuangzi's thoughts concerning the compatibility between being and nonbeing; he also described Daoist cosmological evolution as introduced in a collection of classical essays before 139 BCE, *Huainanzi* (e.g., Liu An, 2010). Jiang and Tang (2002) examined the Daoist theory on the origin of the universe (pp. 187–89). In addition, Chen (1994) explored the Daoist model of cosmological origin and expansion.

Two Aspects of *Dao*

In the Tao Te Ching , Laozi employed the word *dao* as a metaphysical totem. It encapsulates his philosophy of dualistic monism in deciphering the evolution of the universe. To be more specific, *dao* contains both the ontological principle and the cosmological basis.

On the one hand, *dao* has an ontological connotation with both *nothingness* and *beingness*, in contrast to the monism popular in the West (Perkins 2019). The Tao Te Ching describes this dialectic philosophy as follows (Laozi, chapters 1, 40; my translation):

> Nothingness is named the Origination of all under Heaven; Beingness is named the Mother of every entity of reality. . . . All the entities under Heaven are begotten from Beingness, and Beingness is begotten from Nothingness.

Concisely speaking, compared to all the following things (*beingness*), *dao* is *nothingness*; yet, it is not the absolute *nothingness* but the ontological origin of all the things, and thus, it is the *beingness* of all the entities.

On the other hand, complementary to its ontological behavior, the cosmological denotation of the *dao* exposes the essence of the I Ching's *yi*, that is, "change," "becoming," or "evolution." Additional Daoist writings (particularly the three leading classics that follow below) make this point. One of them is the Liezi (written between ca. 475–221 BCE), ascribed to an ancient Chinese philosopher Liezi. It proposes five stages of cosmological evolution (described on p. 43; see also Liezi, chapter 2). The second is the Zhuangzi (written ca. 350–250 BCE), ascribed to an influential philosopher Zhuangzi. This masterpiece inherits the Tao Te Ching's cosmic evolution by articulating that (Zhuangzi, chapter 5; my translation):

> The luminous is born from the obscure, the multiform from the unembodied, the spiritual from the Dao, the bodily from the seminal essence, and, all things from the forms of one another.

It also points out that (Zhuangzi, chapter 11; my translation):

> What has a real existence but has nothing to do with position is the cosmological space, what has a continuance but has nothing to do

> with either beginning or end is the cosmological time . . . all things come from non-existence, but the first existences are unable to bring themselves into existence from non-being except for the non-being which is just the same as the non-existing.

The last text is the Huainanzi (Liu, 139 BCE), which places the cosmological evolution of the *dao* on a more solid ground (He, 1998, pp. 165–66; my translation):

> The *dao* produces a nebulous void; the void produces space-time; the space-time itself gives rise to the primordial *qi* enclosed by a horizon within which the pure-bright part spreads out to give rise to Heaven, and the heavy-turbid one congeals to give rise to Earth.

Primordial *Qi* and Classified *Qi*

Ontologically speaking, the concept of primordial *qi* emphasizes the transcendent, ultimate vital power or energy that, as the cosmic prime mover, initiates and sustains all the aspects of existence by following the cosmic rules of the universe. Different from the term *primordial qi*, the word *qi* is always intricately associated with classified vital power or energy that is dynamic and transformative to permeate any things in the universe and drive all the processes in shaping the cosmological courses of all events.

More or less related to either polytheistic teachings or internal alchemy, various Daoist texts brought together primordial *qi* and specific classified *qi*, and resulting in the formation of *jingqi*, that is, the void *qi* or effulgence *qi*. It was claimed to come from the void and to drive the evolution of entities in reality. According to the Authentic Scripture on the Eight Effulgences of the Primordial Chaos:

- Before heaven and earth . . . there is merely the void where only jingqi resides (appearing as a yellowish hue);
- Jingqi is so tenuous as to become a dim and dusky state when

fluctuating to reach its utmost; when the state fluctuates to an extreme, moistening is nourished;

- The moistness went to form a fog in its extreme; the fog produces water in its extreme; . . . the flow of water gives rise to yinqi;
- At the extreme of yinqi, yangqi is produced;
- The mutual interactions of yinqi and yangqi constitute *hundun* (the primordial chaos) after sufficient times of reciprocal influences;
- The chaos . . . is the mixed *qi* which springs natural things as the mother of Heaven and Earth. (Anonymous, chapter 1; my translation)

Another scripture, the Seven Tablets in the Cloudy Satchel, proposed that there are nine classified kinds of the mixed *qi* in the chaos (Zhang, "Northern Song"; my translation):

- At first, a primordial *qi* emerges from the void;
- After 99,000,000,990,000 years, the produced primordial *qi* transforms three times into three *qi*, respectively, with intervals of the same number of the years in-between;
- The resulting *qi* of these three transformations together produces the supreme *qi* named *wushang*;
- In two intervals with the same number of the years after the supreme *qi* is produced, the second middle *qi* and the third middle *qi* come into being successively;
- The supreme *qi*, the second middle *qi*, and the third one together . . . produce the supremely mysterious *qi* named *xuanlao*;
- In three intervals with the same number of years after the supremely mysterious *qi* is produced, the lower three *qi* are transformed and together . . . produce the grand supreme *qi* named *taishang*.

In total, the period from the chaos of the primordial *qi* to the grand supreme *qi* lasts 891,000,008,910,000 years, that is, about 900 trillion years.

A later scripture, the Explanations and Commentary with Diagrams to the Wondrous Canon of the Eternal Purity and Tranquility as Taught by the Supreme Venerable Sovereign, described the evolution of the chaos in specific stages of *yin* and *yang* (Wang, Song dynasty; my translation):

- initially, a seed of *yang* becomes active (in the chaos) to generate odd and even;
- *yin* and *yang* are thus divided to give birth to heaven and earth;
- the heaven is formed by the light and pure *qi* which rises upward to spread, while the earth is formed by the dense and impure *qi* which solidifies downward to condense;
- yinqi can become active to flow out of the earth and rise up to the heaven, while yangqi can become still to flow out of the heaven and condense down to the earth;
- *yin* and *yang* alternate and circulate endlessly . . . to drive the celestial motion of the Sun and the Moon, the harmonic distribution of the five phases' *qi*, and the proceeding course of the four seasons;
- therefore, all the things are always able to be brought to maturity.

Relation among *Yi*, *Dao*, *Li*, and *Qi*

After 1000 CE, neo-Confucianists established *li-xue*, the learning of *li* (the principle), and *dao-xue*, the learning of *dao*. The principle (*li*) was defined as being the ontological oneness of everything expressed by the previously introduced dialectical monism.

Traditionally, conventional Confucianism was conservative. It emphasized a socially and ethically hierarchical harmony between the emperors and their human subjects, and among the human subjects from one dynasty to the next. By contrast, the neo-Confucian philosophy stressed the correlation and the unity between nature and humanity. It put forward a *li-qi* philosophy that unified *yin* and *yang*. This philosophy rested upon the integration of the ontological *li*, the cosmological *qi*, and the inseparable unity of heaven, earth, and humanity as a whole. It held that, on the one hand, *li* denotes the ontological principle of *yi* in the evolution following *dao*. Put another way, *dao* is the way of *yi*. On the other hand, *qi* represents the cosmological vital power or energy to drive *yi* on the track of *dao*. Thus, *yi* is the change, becoming, or evolution following *dao*. In short, *dao* is nothing else but the "way" of *yi*, while *yi* is nothing else but the change, becoming, or evolution following *dao* (Chen 2011; Zheng 2007).

Moreover, *li* refers to the principle that "all the things originate from *you* (the beingness or existence); and that *you* is begotten of *wu* (the nothingness, voidness, or emptiness)" (Laozi, chapter 40). Yet the word *wu* does not mean *sunyata*, the Buddhist definition of pure emptiness. Instead, it refers to a series of primeval evolving patterns of *yi* that are collectively called *wuji* (the ultimateless) (Makeham, 2010, p. 136), an alternative word to express *dao* (e.g., Zhu, Song dynasty; my translation):

- **Premise 1:** *dao* begets the oneness (Laozi, chapter 1);*
- **Premise 2:** from *wuji*, *taiji* (the great ultimate) comes to be (Zhou 1990, p. 3); and
- **Premise 3:** *taiji* is nothing else but the oneness (section 151 in Zhang, Southern Song).

NEO-CONFUCIAN ONTO-COSMOLOGICAL PHILOSOPHY

Though the *li-qi* philosophy served as the foundation of the neo-Confucianism, it was inherited from Daoist philosophy and thus inseparable from Daoist immanence and totality. Here, "immanence"

*These phrases are shown in Chinese in the compendium, p. 206.

refers to Daoist ontological values and powers in the things themselves; and "totality" refers to the cosmologically interrelated parts of all the things in the cosmological evolution (Mou 2009, p. 93). From the perspective of modern cosmogenesis, the *li-qi* postulates should provide a proper paradigm to account for the evolution of the known universe from its primordial prenatal stage of quantum-gravity fluctuations to a sudden ballooning into the emergence of clumps of matter.

Wuji

As mentioned above, the ontological characterization of *dao* is signified by a series of primeval evolving patterns of *yi* collectively known as *wuji* (Makeham 2010, p. 136). These patterns evolve successively from the primordial state of thoughtlessness, effortlessness, quietness, and stationariness by obeying *li* of the oneness of all the parts, a stage of unity named *taiji* (Sung 1973, p. 295; Pregadio 2008, p. 50).

Before any transformations or transmutations can happen, the very beginning does not possess any primordial *qi*. However, it may develop a state in the void known as *taizhao* (the great inception), a formless fluctuation featuring invisible "ascending and flying, diving and delving" (He 1998, p. 165). This state distributes here and there known as "*xukuo* (the nebulous void) where *Dao* begins to appear" (Dong, 206 BCE, chapter 3; my translation). The void is characterized by several primeval evolving patterns altogether named *wuji* (Liezi, chapter 2; my translation); these are called:

- *taiyi* (the great easiness), at which *qi* is too weak to be seen (though it is produced inside the nebulous void);
- *taichu* (the great origin), at which *qi* emerges (yet, without *xíng*, that is, the shape of matter);
- *taishi* (the great inaugural), at which the shape (xíng) emerges (from *qi*, yet, without *zhì*, that is, the quality of matter);
- *taisu* (the great simplicity), at which the quality (zhì) emerges (from *qi* and xíng, yet, without *tǐ*, that is, the entity of matter).

At this initial stage, "the three ingredients of *qi*, xíng (shape) and zhì (quality) are blended, called hundun (the primordial chaos), where nothing can be separated from anything else, a state invisible to sight, inaudible to hearing, and intangible to touching, thus recognized as *yi*; it possesses neither shape nor bounds and expresses itself as the oneness of all. This oneness is the beginning of the following unprecedented deformations (that is, the distortions or alterations in shape or structure)" (Liezi, chapter 2; my translation). Alternatively, the chaos can be viewed as the unity of three flavors:

- invisible to seeing, thus called the colorless, *yí*;
- inaudible to hearing, thus called the soundless, *xī*; and
- intangible to touching, thus called the shapeless, *wēi*.

> [These] are so inseparable, in no way to be defined independently, that they exist as a mixed-up unity (to express all the patterns of *wuji*). Outside the unity, there is no more lightness; and, inside the unity, there is no more darkness; thus, it is vague enough to defy any description and can be recategorized as nothingness. For this reason, the unity is a form without form, an image of matter without matter, hence it is called *huhuang* (seemingly visible but invisible), of which neither can the front be seen when facing it nor the back be seen when following it. . . . It is the demonstration of the Dao, from which the beginning of the past can be known (Laozi, chapter 14; my translation).

Such a unity consists of all the primeval evolving patterns, named *taiji*, as defined earlier. It has the pregnant potential as the origin of all the entities of nature. It should be noted here that, in the arithmetic system, *wuji* (or *dao*) is equivalent to number zero (0), and *taiji* to the number one (1) (Hu and Li 2006).

Taiji

Figure 3.1 illustrates the structure of *taiji*. It is made up of two components: *yin* and *yang*. Both of them always coexist in the *yin-yang* pair to

Figure 3.1. *Taiji* diagram

demonstrate the two contradictory but complementary features inherent in all the entities of nature: that which is substance, activity, and dominance, or the positive, masculine, versus that which is function, inertia, and subordination, or the negative, feminine.

Within the outermost circle of figure 3.1, the active nature of the oneness gives rise to the *yang* category; yet, at the extreme of its activity, the *yang* property becomes tranquil and then falls into the tranquilness of the *yin* category; at the extreme of the tranquility, the *yin* property becomes active, and the activity generates *yang* again; activity and tranquility are therefore alternating, and each is the root of the other; the *yin* and *yang* properties are thus distinguished and sustained as *liangyi*, that is, the two polarities of taiji (Zhou 1990, p. 3).

In addition, figure 3.1 reveals that *yin* and *yang* are opposite yet complementary. They form a pair of spiral fishlike shapes, head to tail against each other, to cooperatively drive a couple of processes (Zhou 1990, p. 26; my translation):

- either being *yang* in activity instead of *yin* in tranquility, or being *yin* instead of *yang*, forms *wùxìng* (the thingness* of things), and

**Thingness* refers to "the quality or state of objective existence or reality" ("Thingness," 2019, def. 1), in contrast to *objecthood*, which is the "condition or state of being an object" ("Objecthood," 2019, def. 1).

- either being active but with a lack of *yang*, or being tranquil but with a lack of *yin*, forms *shénxìng* (the unfathomability* of things)

However, "both being active but lacking *yang* and being tranquil but lacking *yin* do not mean being fully devoid of *yang* and *yin*, respectively," because "there always exists tranquility within activeness and activeness within tranquility" (Zhou 1990, p. 26; my translation). Though *yin* and *yang* are opposite in polarity and interdependent of each other, there is *yang* in *yin*, and there is *yin* in *yang*, and, *yin* and *yang* always remain complementary and mutually transformed and consumed in the unity of both: taiji. The continual interplay and transformation between the correlative *yin* and *yang* constantly fuel the flows of *qi*. The resultant trend of either *yin* or *yang* takes a spiral fishlike form, both of which are structured like a pair of fishes nestling head to tail against each other (Robinet 2008, p. 934).

Consequently, the two polarities of taiji, *yin* and *yang*, are the two correlative elements, named *liangyi*, evolving from wuji responsible for any cosmological evolutions. In summary, the general characteristics of them are as follows (Maciocia 2015, p. 7). They:

- behave interdependently with each other;
- are opposite in polarity;
- consume each other;
- transform mutually into each other; and
- form an integrated and inseparable body.

What is more, "*dao* resides in *yin* and acts in *yang*" (Zhu Xi 1992, p. 141; my translation). On the one hand, in view of the infrastructural

*Relative to *wùxìng* (the thingness), there is a concept of *shénxìng* (the unfathomability). While the former refers to the "quality or state" of the objective existence or reality, the latter refers to the "rationality" of the objective existence or reality. This "rationality" is "impossible to be comprehended or understood," that is, it is unfathomable. In the I Ching, "the unpredictability of *yin* and *yang* is called unfathomability" (e.g., Sung, 1973, p. 281).

quality, "to sustain *dao* . . . is the duty of *yang* in the accomplishment of transformation and cultivation, while to achieve *dao* . . . is the duty of *yin* in the nourishment of promotion and diversity." On the other hand, in view of the exterior behavior of phenomena, "endless alternative interchanging and generation is called *yi*; *yin* generates *yang* and *yang* generates *yin* in myriad transformations." Furthermore, "when *yin* (or *yang*) dominates, it serves as the functioning polarity of the transformation; and, when it does not dominate, it provides the condition for the other to behave as the functioning polarity of the transformation" (Laozi, chapter 11; my translation).

Yi of *Yin* and *Yang*

The mutual responsiveness of taiji's two polarities, *yin* and *yang*, triggers further correlative cause-effect interactions of *yi*. The interactions are best described in the I Ching (Sung 1973, p. 299). The process starts from the formation of taiji's arithmetical signature, 1. Beyond that:

- the 1's "liangyi produces 4 *xiang* (*sixiang*: four digrams)";
- then, "4 *xiang* produces 8 *gua* (*bagua*: eight trigrams)";
- afterward, the "mutual communion and alternation between the resolute (*yang*) and the yielding (*yin*), together with the full combinations of the two 8-gua sets," construct 64 *gua* (*liushisigua*: sixty-four hexagrams). (Legge 1963, p. 348; Sung 1973, p. 272)

In this way, *yin* and *yang* are changing in ascending numerological order into a series of hierarchical systems. In this system, *qi* is distributed and redistributed again and again "from the most pure, rational category at first to the next realistic, perceptive category . . . of the phenomenal world . . . by means of the abstract thinking" on such things like location, element, season, color (Hegel 1983, pp. 122–23).

Consequently, *yin* and *yang* present a binary change, *yi*, starting from the ontological origin, zero (0), i.e., wuji. In the numerological expression, a single line is known as a *monogram*. If the line is broken, it is designated as representing *yin*; if the line is solid, it is designated as representing *yang*. Thus, a monogram can express these two polarities. To elaborate further, two monograms can be combined to symbolize four digrams, and so on. In general, *yi* evolves following this chain: wuji (0; primeval) → taiji ($1 = 2^0$; 0-D) → two monograms ($2 = 2^1$; 1-D) → four digrams ($4 = 2^2$; 2-D) → eight trigrams ($8 = 2^3$; 3-D) → sixty-four hexagrams ($64 = 8 \times 8 = 2^6$; two independent but interlocking 3-D).

Two Monograms

A monogram has two possibilities in its polarity, positive or negative. Two polarized monograms are used to designate *yang* and *yin*, respectively, named *yangyao* (positive line) and *yinyao* (negative line).

Yangyao is the masculine line and signifies *yang*. It is symbolized by a solid line, — , expressing such characteristics as, but is not limited to, positive polarity, heaven, male, resoluteness, activity, energy, and the principle of unity.

Yinyao is the feminine line and signifies *yin*. It is symbolized by a broken line, -- , expressing such characteristics as, but is not limited to, negative polarity, earth, female, yielding, tranquility, matter, and the principle of duality.

Four Digrams and Five Phases

The four digrams come from the four correlative combinations of the above two polarized monograms, — and -- :

young yang ⚎, corresponding to the phase wood;
mature yang ⚌, corresponding to the phase fire;
young yin ⚍, corresponding to the phase metal;
mature yin ⚏, corresponding to the phase water.

An alternative description of the four digrams is *wuxing* (five phases). These phases occurred in the ancient Chinese astronomical observations from as early as 2000 BCE. They were used together in celestial records of the Sun, the Moon, and the five visible planets: water planet (Mercury), metal planet (Venus), fire planet (Mars), wood planet (Jupiter), and earth planet (Saturn). These planets were observed to all be conjunct three times, including: (*a*) on February 26, 1953 BCE, when they were seen in a conjunction spanning an arc of 4° in the sky; (*b*) on December 20, 1576 BCE, when they again appeared within a 4° arc with the exception of Venus (which was within 41°–45°); and (*c*) on May 28, 1059 BCE, when they were in an arc of 7° (Pankenier 1981, 1983, 1998, pp. 161–95).

In the *li-qi* category, the five phases contain all those represented by the four digrams above, plus an extra element, earth. This phase is the center of the five phases and is called the *tian-run* (heavenly nourishment) (Dong, 206 BCE). Note that this phase has no relation with the prenatal taiji which is originated from wuji. Instead, it behaves as the postnatal taiji correlated with the evolved two polarities and the four digrams. At this phase, distinct entities emerge from matter after the prior stage of *taisu* (the great simplicity), at which, as mentioned previously, matter comes from *qi* and form, yet without distinct entities (Xu 2002, p. 2). Table 3.1 lists the names and the properties of all the four digrams and related five phases. Phase earth resides at the central palace (no. 3), surrounded by the four digrams (nos. 1, 2, 4, 5) in the following order: wood, fire, earth, metal, and water. Wood is the beginning of all, and water is the end.

TABLE 3.1. FOUR DIGRAMS AND FIVE PHASES

No.	Digram	Name	Direction	Dynamic	Season	Color	Animal
1	⚎	Young Yang	East	Wood	Spring	Azure	Dragon
2	⚌	Mature Yang	South	Fire	Summer	Red	Bird
3	☯	Heavenly Nourishment	Center	Earth	Late Summer	Yellow	–
4	⚍	Young Yin	West	Metal	Autumn	White	Tiger
5	⚏	Mature Yin	North	Water	Winter	Black	Turtle

Eight Trigrams

The eight trigrams, also 8 gua (or bagua), come from the eight possible combinations of the two polarized monograms, — and -- , and the four digrams, ⚍, ⚌, ⚎, and ⚏. Table 3.2 lists the names and the properties of all the eight trigrams. There exist a couple of 8-gua sets: Fuxi's prenatal or noumenal one, and King Wen's postnatal or phenomenal one. The former corresponds to a set of compass locations written as Location-1 in table 3.2 (Zhu Xi 1992, p. 170); while the latter corresponds to a set of the ones written as Location-2 in the table (Sung 1973, p. 343). The properties of the eight trigrams at Location-2 are correlated with both the five phases and eight solar terms (Chen 2015, p. 326).

TABLE 3.2. EIGHT TRIGRAMS

No.	Trigram	Name	Sign	Location-1	Location-2	Emblem	Solar Term
1	☰	Qian	Heaven	South	Northwest	Metal	Winter Beginning (WB)
2	☱	Dui	Lake	Southeast	West		Autumn Equinox (AE)
3	☲	Li	Fire	East	South	Fire	Summer Solstice (SS)
4	☳	Zhen	Thunder	Northeast	East	Wood	Spring Equinox (SE)
5	☴	Xun	Wind	Southwest	Southeast		Summer Beginning (SuB)
6	☵	Kan	Water	West	North	Water	Winter Solstice (WS)
7	☶	Gen	Mountain	Northwest	Northeast	Earth	Spring Beginning (SpB)
8	☷	Kun	Earth	North	Southwest		Autumn Beginning (AB)

Sixty-Four Hexagrams

The sixty-four hexagrams, also 64 gua (or liushisigua), come from the permutations of the pairwise combinations of the two sets of the eight trigrams. Similar to eight trigrams, there are two types of the 64-gua sequences as given in table 3.3. The table lists the full names and definitions of all the 64 hexagrams in King Wen's postnatal sequence (with hexagram numbers in **black**), accompanied by Fuxi's prenatal sequence (with binary codes in red).

TABLE 3.3. SIXTY-FOUR HEXAGRAMS

King Wen (No. in **Black**) and Fuxi (No. in Red)

Upper Jing (1–30)			Lower Jing (31–64)		
No.	**Name**	**Definition**	**No.**	**Name**	**Definition**
1/63	䷀ Qian	Pure yang / Heaven / Creativity	*31/14	䷞ Xian	Response
2/0	䷁ Kun	Pure yin / Earth / Receptivity	32/28	䷟ Heng	Duration
3/34	䷂ Zhun	Initial difficulty	33/15	䷠ Dun	Retreat
4/17	䷃ Meng	Ignorance	34/60	䷡ Dazhuang	Great strength
5/58	䷄ Xv	Waiting	35/5	䷢ Jin	Going upward
6/23	䷅ Song	Litigation	36/40	䷣ Mingyi	Injured brightness
7/16	䷆ Shi	Army	37/43	䷤ Jiaren	Family members
8/2	䷇ Bi	Assistance	*38/53	䷥ Kui	Opposition
9/59	䷈ Xiaoxu	Small accumulation	39/10	䷦ Jian	Obstruction
*10/55	䷉ Lv	Treading	40/20	䷧ Xie	Deliverance
11/56	䷊ Tai	Peace / Tranquility	*41/49	䷨ Sun	Decrease
12/7	䷋ Pi	Adversity	42/35	䷩ Yii	Increase
13/47	䷌ Tongren	Fellowship	*43/62	䷪ Guai	Resoluteness
14/61	䷍ Dayou	Great possession	44/31	䷫ Gou	Meeting
15/8	䷎ Qiann	Humility	*45/6	䷬ Cui	Gathering
16/4	䷏ Yv	Enjoyment	46/24	䷭ Sheng	Ascending
*17/38	䷐ Sui	Following	*47/22	䷮ Kunn	Besetment
18/25	䷑ Gu	Work on decay	48/26	䷯ Jing	The well
*19/48	䷒ Lin	Approaching / Overseeing	*49/46	䷰ Ge	Revolution
20/3	䷓ Guan	Contemplation / Correspondence	50/29	䷱ Ding	The cauldron
21/37	䷔ Shihe	Biting to close up	51/36	䷲ Zhen	Shaking

*These are the 15 hexagrams which Laozi relied upon to write the Tao Te Ching. According to the I Ching, the 64 hexagrams are divided into two parts, the *upper jing* and the *lower jing*. The former contains 30 hexagrams, and the latter contains 34. While King Wen's sequence was given in the I Ching (e.g., Legge 1891, pp. 257–64), Fuxi's sequence was used in the neo-Confucian numerological cosmology (Shao 1993, Song-a, Song-b).

TABLE 3.3. SIXTY-FOUR HEXAGRAMS

King Wen (No. in **Black**) and Fuxi (No. in Red) (*cont.*)

Upper Jing (1–30)			Lower Jing (31–64)		
No.	**Name**	**Definition**	**No.**	**Name**	**Definition**
22/41	Bii	Decoration	52/9	Gen	Stopping
23/1	Bo	Disintegration	53/11	Jiann	Gradual development
24/32	Fu	Returning	*54/52	Guimei	Young sister to marry
25/39	Wuwang	Non-fault	55/44	Feng	Abundance
26/57	Daxu	Great accumulation	56/13	Lvv	Wandering
27/33	Yi	Nourishment	57/27	Xun	Gentleness
*28/30	Daguo	Great passing	*58/54	Dui	Joy
29/18	Kan	Repeated sinking	59/19	Huan	Dispersion
30/45	Li	Brightness	*60/50	Jie	Restriction
			*61/51	Zhongfu	Inner sincerity
			62/12	Xiaoguo	Small passing
			63/42	Jiji	Completion
			64/21	Weiji	Non-completion

*These are the 15 hexagrams which Laozi relied upon to write the Tao Te Ching. According to the I Ching, the 64 hexagrams are divided into two parts, the *upper jing* and the *lower jing*. The former contains 30 hexagrams, and the latter contains 34. While King Wen's sequence was given in the I Ching (e.g., Legge 1891, pp. 257–64), Fuxi's sequence was used in the neo-Confucian numerological cosmology (Shao 1993, Song-a, Song-b).

For comparison, table 3.4 gives Fuxi's binary codes (numbers in red) accompanied, respectively, by King Wen's numbers (in black). The two notations, "Up" and "Down," indicate the upper and lower eight trigrams, respectively, of all 64 hexagrams. In table 3.4, there are 8 hexagrams in the yellow-shaded region, called *jinggua*, that is, the superimposed hexagrams. They are the superimposition of two identical trigrams upon each other. They occupy the primary diagonal codes from the bottom-right corner to the top-left one. They have the red-colored Fuxi's codes in an arithmetic sequence by a factor of 9. This number is the top number in the I Ching to denote the highest *yang*. The superimposition is as follows:

TABLE 3.4. SEQUENCE OF SIXTY-FOUR HEXAGRAMS

Fuxi (No. in Red) and King Wen (No. in **Black**)

Up / Down	☷	☶	☵	☴	☳	☲	☱	☰
☷	0/2	1/23	2/8	3/20	4/16	5/35	6/45	7/12
☶	8/15	9/52	10/39	11/53	12/62	13/56	14/31	15/33
☵	16/7	17/4	18/29	19/59	20/40	21/64	22/47	23/6
☴	24/46	25/18	26/48	27/57	28/32	29/50	30/28	31/44
☳	32/24	33/27	34/3	35/42	36/51	37/21	38/17	39/25
☲	40/36	41/22	42/63	43/37	44/55	45/30	46/49	47/13
☱	48/19	49/41	50/60	51/61	52/54	53/38	54/58	55/10
☰	56/11	57/26	58/5	59/9	60/34	61/14	62/43	63/1

No.	Trigram	Superimposed trigram	Hexagram jinggua
1	☰	䷀	1/63
2	☱	䷹	58/54
3	☲	䷝	30/45
4	☳	䷲	51/36
5	☴	䷸	57/27
6	☵	䷜	29/18
7	☶	䷳	52/9
8	☷	䷁	2/0

Yi of *Li* and *Qi*

Yi of *li* and *qi* presents the genesis of natural things in reality. It follows taiji's *li* in addition to its commitment to the numero-cosmological evolution of *yin* and *yang* (Mou 2009, pp. 71–72). A snapshot of the paradigm is given by figure 3.2 (p. 54).

The *li-qi* model tells us that everything obeys a sequential evolution owing to its self-embodiment of both the ontological *li* and the cosmological *qi*. It describes four themes as follows, as discussed previously in "Neo-Confucian Onto-Cosmological Philosophy" (p. 42 and illustrated in figure 3.2).

First, as described in the subsection "Wuji" (p. 43) and shown in the right-hand-side of figure 3.2 at the stage of wuji (the ultimateless), *dao* begins from xukuo (the nebulous void) in taizhao (the Great Inception), *qi* emerges in taichu (the Great Origin) after accumulating in taiyi (the

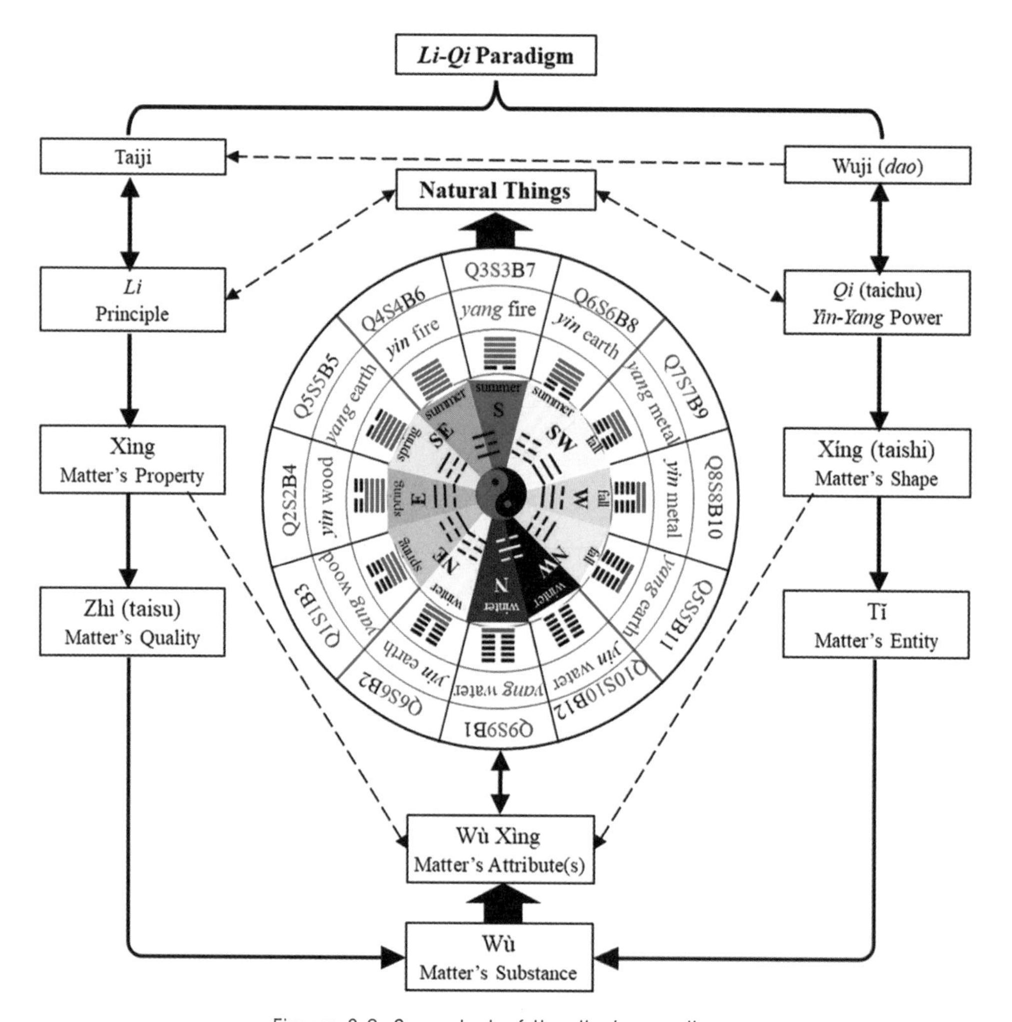

Figure 3.2. Snapshot of the *li-qi* paradigm

Great Easiness), matter's shape (xíng) emerges from *qi* in taishi (the great inaugural), and matter's quality (zhì) emerges from *qi* and xíng.

Second, as described also in the "Wuji" subsection and shown in the left-hand-side part of figure 3.2, wuji's three ingredients (*qi*, xíng, zhì) are so blended to behave as the primordial chaos (hundun), which follows the principle (*li*) of the oneness, known as taiji (the great ultimate), that determines matter's property (xìng) from which matter's entity (tǐ) is subsequently developed.

Third, as shown in the bottom part of figure 3.2, matter's quality (zhì) and entity (tǐ) are so integrated together to become the antecedent source in the production of matter's substance (wù) which owns diverse attributes of matter (wù xìng) to differentiate itself from others (c.f. Zheng, 2007).

Finally, as shown in the middle circle of figure 3.2 (in accordance with table 3.5), matter's attributes (wù xìng) are ubiquitously described by wuxing (the five phases), the symbolic system used to represent the cosmological categories, characteristics, interactions, or relationships among natural things. In figure 3.2, the elements of wuxing are associated with symbolic colors, both heavy to represent *yang* and light to *yin* (wood in green; fire in red; earth in yellow; metal in white; and water in black). Each of the five has *yin* and *yang*, respectively, giving 10 types of *yin-ying*'s *qi* as labeled in dark red from Q1 to Q10 (yang wood, yin wood; yang fire, yin fire; yang earth, yin earth; yang metal, yin metal; and yang water, yin water). Notice that phase earth is different from the other four elements of wuxing (wood, fire, metal, water). While these four elements have only one *yin* and one *yang*, it has two *yin* and two *yang* (Q5S5B5 *yang*, Q5S5B11 *yang*, Q6S6B8 *yin*, and Q6S6B2 *yin*) to participate in balancing the distributions of *qi*. All these types of *qi* in turn are correlated with all other sets of cosmoecological elements (see table 3.5 below), including:

- A set of eight directions: east (**E**), south (**S**), west (**W**), north (**N**), northeast (**NE**), southeast (**SE**), northwest (**NW**), and southwest (**SW**);

- A set of four seasons: spring, summer, fall, and winter;
- A set of eight trigrams: T1-*qian* ☰, T2-*dui* ☱, T3-*li* ☲, T4-*zhen* ☳, T5-*xun* ☴, T6-*kan* ☵, T7-*gen* ☶, and T8-*kun* ☷;
- A set of ten celestial stems: S1-*jia*, S2-*yi*, S3-*bing*, S4-*ding*, S5-*wu*, S6-*ji*, S7-*geng*, S8-*xin*, S9-*ren*, and S10-*gui* (Nielsen 2003, pp. 9–10; Xiao 2014, p. 13);
- A set of twelve terrestrial branches (Mei 1739): B1-*zi*, B2-*chou*, B3-*yin*, B4-*mao*, B5-*chen*, B6-*si*, B7-*wu*, B8-*wei*, B9-*shen*, B10-*you*, B11-*xv*, and B12-*hai*; and
- A set of twelve sovereign hexagrams (*zhugua*), that is, twelve waning-waxing (*xiaoxi*) hexagrams: H1-*fu* 24/32 ䷗, H2-*lin* 19/48 ䷒, H3-*tai* 11/56 ䷊, H4-*dazhuang* 34/60 ䷡, H5-*guai* 43/62 ䷪, H6-*qian* 1/63 ䷀, H7-*gou* 44/31 ䷫, H8-*dun* 33/15 ䷠, H9-*pi* 12/7 ䷋, H10-*guan* 20/3 ䷓, H11-*bo* 23/1 ䷖, and H12-*kun* 2/0 ䷁ (Nielsen 2003, p. 275).

TABLE 3.5. CORRELATION OF QI WITH COSMOECOLOGICAL ELEMENTS

Qi	Q1	Q2	Q3	Q4	Q5		Q6		Q7	Q8	Q9	Q10
	yang	*yin*	*yang*	*yin*	*yang*		*yin*		*yang*	*yin*	*yang*	*yin*
Five Phase	wood		fire		earth				metal		water	
Direction	NE	E	S	SE		NW	NE	SW		W	N	NW
Season	spring		summer		spring	fall	winter	summer	fall		winter	
Eight Trigram	T4 ☳	T5 ☴	T3 ☲		T7 ☶		T8 ☷		T1 ☰	T2 ☱	T6 ☵	
Celestial Stem	S1	S2	S3	S4	S5		S6		S7	S8	S9	S10
Terrestrial Branch	B3	B4	B7	B6	B5	B11	B2	B8	B9	B10	B1	B12
Sovereign Hexagram	H3	H4	H7	H6	H5	H11	H2	H8	H9	H10	H1	H12
	䷊	䷡	䷫	䷀	䷪	䷖	䷒	䷠	䷋	䷓	䷗	䷁
	11/56	34/60	44/31	1/63	43/62	23/1	19/48	33/15	12/7	20/3	24/32	2/0

Square-Circular Diagram

The *li-qi* model attains its onto-cosmological climax with the 64 × 64 square-circular diagram (*fangyuantu*). The diagram demonstrates two main notions: (1) the ontological essence, *li*, of the ultimate perceptible reality; and (2) the cosmological evolution propelled by *qi* within the framework of the human mind as it perceives the known universe (Ding 2005). Figure 3.3 shows the illustration.

This diagram contains three parts to symbolize the circular heaven, the square earth, and the human realm, respectively:

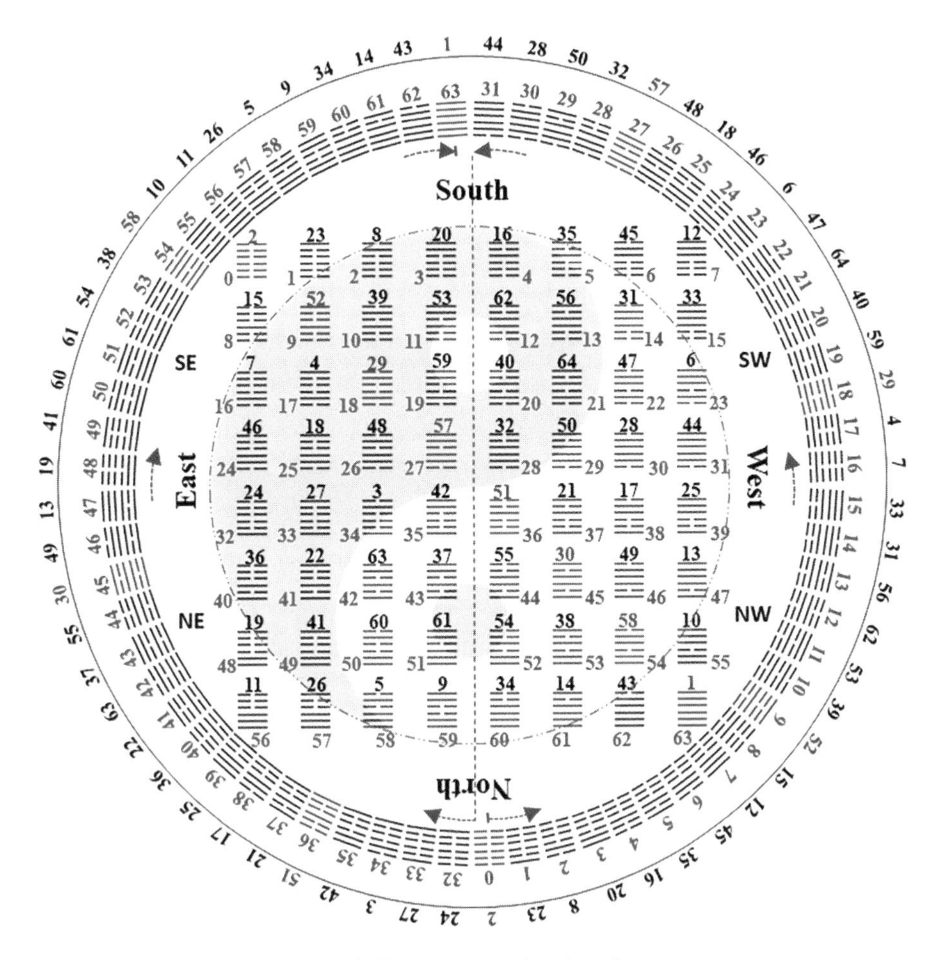

Figure 3.3. The square-circular diagram (adapted from Shao 1993, foreword, figure 3)

- the external circular part is a set of the 64 hexagrams, enclosing the square part inside, to represent the heavenly reality;
- the internal square part is an 8 × 8 matrix of another set of the 64 hexagrams, dwelling in the central region, to represent the earthly reality;
- the space between the circle and the square represents the room where the human realm goes through its transformations. (Major 1993, pp. 32–35)

All the numbers in black and red attached to the elements of the matrix and the circle refer to King Wen's numbers and Fuxi's codes of the 64 hexagrams, respectively, as presented in tables 3.3 and 3.4 (p. 52 and 53). The matrix provides the locations and directions to describe spatial features, while the circle represents the evolutionary patterns that exhibit temporal features.

Both the matrix and the circle describe the rheological characterizations of space and time. The hexagrams in the matrix correspond to the ordered changes in space of the 64 hexagrams row by row and column by column, while those in the circle include four principal hexagrams (*zhenggua*) and 60 ruling hexagrams (*zhigua*) which form the continuous flow of time. These features are discussed in what follows. For convenience, let us pay attention to the hexagram numbers in red in figure 3.3.

The square matrix appears to be nothing else but that given in table 3.4. Colored in purple are the reversely ordered eight superimposed hexagrams (jinggua), lying in the primary diagonal line from southeast to northwest. In the matrix, the eight trigrams, ☰, ☱, ☲, ☳, ☴, ☵, ☶, and ☷, are distributed to occupy all the upper parts of the 64 hexagrams, one of each in every column from right to left; at the same time, they fill all the lower parts of the eight rows, one of each in every row from bottom to top. Note that the 64 Fuxi codes in the matrix increase continuously from 0 to 63, column after column, and row after row.

In addition, the matrix can be divided into four blocks, as labeled within the circle in figure 3.3: the southeast (SE), the southwest (SW), the northeast (NE), and the northwest (NW), with sixteen hexagrams each. The two diagonal ones (i.e., SE to NW and SW to NE) are mirror images. However, the single lines with the hexagrams display opposite polarities.

Moreover, from 0 to 63, the annulus has a horizontal S-shaped sequence of the red Fuxi codes, as labeled by the arrows: the first 32 hexagrams from 0 to 31 are ordered counterclockwise to give the RHS half circle; and the remaining ones from 32 to 63 follow a clockwise pattern to form the LHS half circle. More importantly, the continuous increment of the red codes demonstrates that the counter-clockwise, RHS half circle corresponds to the upper four rows of the matrix, while the clockwise, LHS half circle corresponds to the lower four rows of the matrix. The entire trend of the code sequence turns out to expose the spiral fishlike taiji diagram given in figure 3.1 (p. 45).

HOLOGRAPHY OF WORLD-ORDERING CYCLES

The square-circular diagram presented in figure 3.3 highlights an atheistic and cosmogonic *li-qi* system. It aggregates and surpasses the I Ching's numerological interrelations through illustrating the correlations among the circular heaven, the square earth, and the transforming humanity residing in the free space between the first two (Teiser 2002, pp. 3–37). Based on the correlative integration of the five phases, the diagram was holographic, applicable to the human bodily cycle of 24 hours, the micro-cosmic lunisolar cycle of 24 solar terms (Nielsen 2003, pp. 75–76), as well as the mesocosmic ice-age cycle of 24 world-ordering terms.

Correlative Five Phases

The conceptual cosmoecological scheme illustrates how the five phases (wood, fire, earth, metal, and water) are the carriers of *yin* and *yang* by interacting and correlating with each other. In general, they do so in three dominant ways (Dong 206 BCE; Maciocia 2015, pp. 27–28):

- Generation (*sheng*), that is, a "generating" process of a phase to produce another adjacent phase cyclically in the order of the five phases, such as a mother giving birth to her child;
- Conquest (*ke*), that is, a "conquering" process of a phase through restricting or inhibiting the phase following the next one, such as a grandmother controlling her grandchild;
- Insult (*wu*), that is, an "overreacting" process of a phase through reversely controlling the phase before the last one, such as a grandson insulting his grandmother.

Figure 3.4 depicts the interaction and correlation of these phases. In the figure, the "generating," "conquering," and "insulting" processes are indicated by the boundary-free arrows, the solid-boundary ones, and the dash-boundary ones, respectively. Specifically,

- the "generating" one follows the cycle of wood → fire → earth → metal → water → wood;
- the "conquering" one follows the cycle of wood → earth → water → fire → metal → wood; and
- the "insulting" one follows the cycle of wood → metal → fire → water → earth → wood. (Liu and Liu 2010, pp. 21–27)

As given by table 3.5 (p. 56), *yin-yang*'s *qi* owns 10 components which are correlated with both 10 trigram-designated celestial stems and 12 hexagram-designated terrestrial branches. Such an interlocked celestial-terrestrial orderliness contributes to an onto-cosmological holism. This orderliness captures the essence of cosmoecology, elucidating the interrelations and the interactions in a wide array of realities of different scales. Thus, the holism brings up an enhanced interpretation of the theme related to the holographic unity of heaven, earth, and humanity.

To attain the goal, we need to reconsider the relationship between the "outside" universe and our "inside" selves: in reality, we have never been independent of the universe; instead, our existence has always been

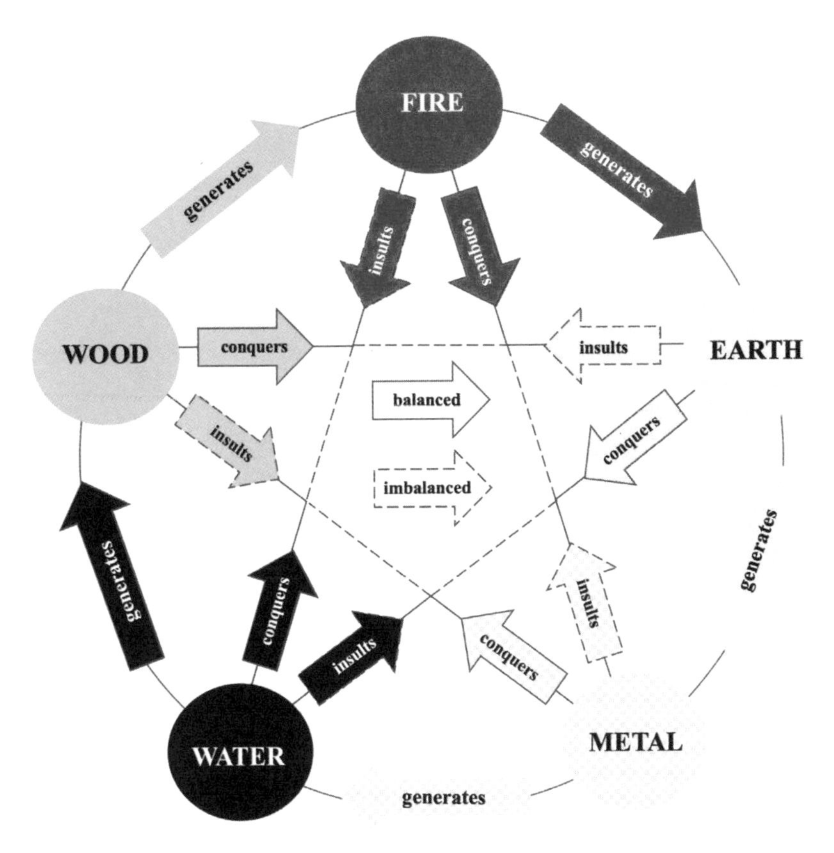

Figure 3.4. Interaction and correlation of the five phases (data adapted from Northern Shaolin 2019)

an integrated part of it, living in the natural and social environments evolved from it holographically (cf. Matthews 2017, p. 31).

Human Bodily *Qi* Cycle

A human body is a miniature microcosm where the holistic effects of feng shui are present and knowable. The biological cycle of the body reflects the functions of a group of twelve internal organs (in order from O1 to O12): gall bladder (*dan*; GB); liver (*gan*; LV); lung (*fei*; LU); large intestine (*dachang*; LI); stomach (*wei*; ST); spleen (*pi*; SP); heart (*xin*; HT); small intestine (*xiaochang*; SI); urinary bladder (*pang-guang*; UB); kidney (*shen*; KI); pericardium (*xinbao*; PC); and the three burners (*sanjiao*; TB)* (e.g., Z. Liu and Liu, 2010, pp. 21–27). Driven by both *yin* and *yang*, these organs take turns being active to sustain life. Table 3.6 gives the correlations among *qi*, the five phases, and the hours in the day when the organs are most active.

TABLE 3.6. HUMAN BODILY *QI* CYCLE

Organ		Qi		Five Phase	Direction	Eight Trigram	Celestial Stem	Terrestrial Branch	Sovereign Hexagram			Hour of Qi Dominance
01	GB	Q9	*yang*	water	N	T6	S9	B1	H1		32	23–1
02	LV	Q6	*yin*	earth	NE	T8	S6	B2	H2		48	1–3
03	LU	Q1	*yang*	wood		T4	S1	B3	H3		56	3–5
04	LI	Q2	*yin*		E	T5	S2	B4	H4		60	5–7
05	ST	Q5	*yang*	earth	SE	T7	S5	B5	H5		62	7–9
06	SP	Q4	*yin*	fire		T3	S4	B6	H6		63	9–11
07	HT	Q3	*yang*		S		S3	B7	H7		31	11–13
08	SI	Q6	*yin*	earth	SW	T8	S6	B8	H8		15	13–15
09	UB	Q7	*yang*	metal		T1	S7	B9	H9		7	15–17
010	KI	Q8	*yin*		W	T2	S8	B10	H10		3	17–19
011	PC	Q5	*yang*	earth	NW	T7	S5	B11	H11		1	19–21
012	TB	Q10	*yin*	water		T6	S10	B12	H12		0	21–23

There are four types of *qi* in a human body: *yuan-qi* (the inborn *qi*, or the source *qi*)†, *zong-qi* (the pectoral *qi*), *ying-qi* (the nutrient *qi*), and *wei-qi* (the protective *qi*).

All of them run through the system of meridians and submeridians that are the paths of body channels and organ collaterals. Every two hours, *qi* moves from one organ to the next. Whenever the balance

*Definition of *sanjiao* (three burners, TB): a system of the body's three "cavities" in charge of the movement, transformation, and excretion of body fluids, including the chest cavity (upper burner), the abdominal cavity (middle burner), and the pelvic cavity (lower burner). See, for example, Maciocia (2015, p. 135) for details.

†*Yuan-qi* (the inborn *qi*, or the source *qi*) is assumed as body's *yuanyi* (the primordial *qi*).

between *yin* and *yang* breaks up, an organ becomes disordered and suffers illness (Maciocia 2015, pp. 43–74). Figure 3.5 illustrates the daily circulation of *qi* inside and on the surface of the human body.

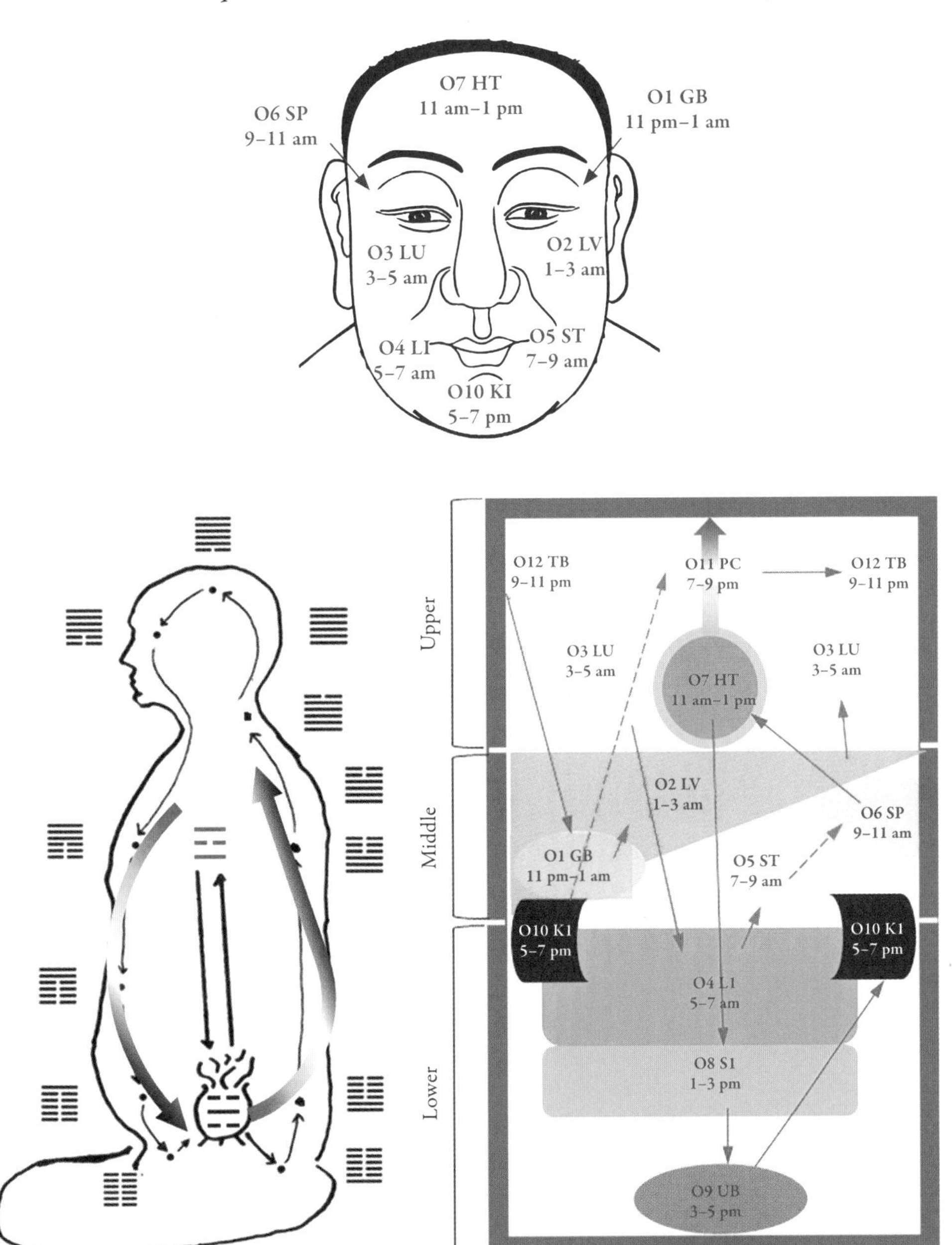

Figure 3.5. Daily circulation of *qi* outside and inside a human body

Qi passes through the organs in a complete cycle of twenty-four hours. Every organ's *qi* dominates a period of two hours (*shichen*), as shown in the last column of table 3.6. During that period, the designated organ is at the maximal cause-effect correlation with its external conditions. At the same time, all the organs are responsible for regularly supplying sufficient *qi* via the meridians for the needs of the body's three treasures, that is, *jing* (essence), *qi*, and *shen* (spirit), as well as additional nutritious needs (e.g., sweat, blood, sinew), in sustaining life and maintaining optimal health. The circulation of *qi* obeys the chain of the twelve sovereign hexagrams from H1 to H12.

The first circulating moment falls precisely at high midnight during the presiding terrestrial branch, B1-*zi* (23-1). During this period, O1-GB needs to be protected the most. The related sovereign hexagram is H1-䷗32. At midnight, *yang* begins to rise in the body, in sharp contrast to that of the *yin*, which begins to rise at high noon (12:00) during the presiding terrestrial branch, B7-*wu* (11-13), marked by the related sovereign hexagram, H7-䷫31. While wakefulness is harmful in the middle of the night (B1-*zi*) to the gall bladder (GB), excessive activity is harmful at noon (B7-*wu*) to the heart (HT).

Microcosmic Lunisolar Calendrical System

Beyond the miniature microcosm of a human body lies a larger cycle of the microcosmic lunisolar calendrical system. It describes regular natural events, such as the seasons. The cycle correlates with the *yin-yang* changes related to the five phases and can be used to predict possible mishaps and catastrophes (Li 2002; Liang 2007, p. 3). Figure 3.6 gives a snapshot of this cycle by taking year 2024 as an example.

The center of the cycle is the phase earth. Both the lunisolar and the Gregorian calendrical systems are used in the figure. The former is useful for agricultural applications, and the latter applies to daily use worldwide, as well as for the modern big-bang cosmology discussed later. The cycle displays harmonic and balanced correlations among the following sets of the cosmo-ecological elements:

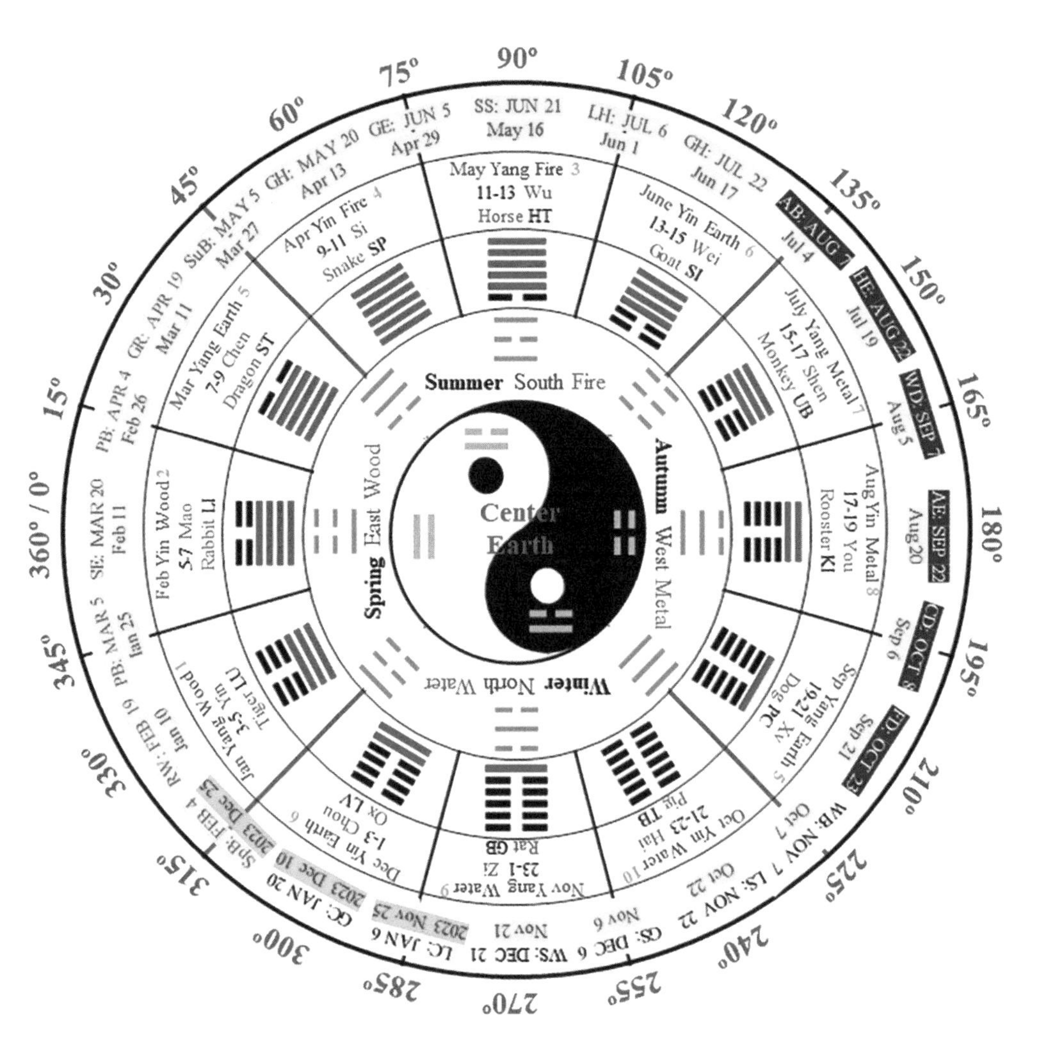

Figure 3.6. Calendrical system showing both lunisolar and Gregorian correlations (2024). As an example, the abbreviation Nov refers to the lunisolar November, while NOV refers to the Gregorian November.

- two monograms that symbolize the lunisolar *yin* and *yang*;
- four digrams that symbolize the four seasons;
- eight trigrams that symbolize the eight directions (four are labeled);
- twelve sovereign hexagrams that refer, respectively, to the 12 organs of the body, the 12 zodiacal animals, the 12 two-hour slots

of activity, the 12 terrestrial branches, and the 12 lunisolar months (taking year 2024 as an example), as well as the 10 celestial stems; and
- the 24 solar terms and the corresponding 24 solar longitudes along the ecliptic, given in Gregorian months and on Gregorian days.

The history of the 24 solar terms dates back to the time when Fuxi invented eight trigrams and then set up the earliest calendrical system, the fire-star calendar, also known as the alpha-scorpii calendar (Zhou 2017, sec. 97; Yang 2008, pp. 247–48). It appeared in the first Chinese dynasty, the Xia dynasty (ca. 2205–1766 BCE). It was thus named *xiali* (the Xia Calendar). During the second dynasty, the Shang dynasty (1523–1027 BCE), the Moon and the Sun were expressed by the signs — and --, respectively, seen in the ancient waterclocks and sundials (York 1997, pp. 16–21). With the help of the symbols, it was easy to record about 365 days of a solar year and about 30 days of a lunar month. The signs also helped to fix a set of four solar terms: J4 spring equinox (SE), J10 summer solstice (SS), J16 autumn equinox (AE), and J22 winter solstice (WS) (Mark 2016).

In the period of the Warring States, four additional solar terms were identified and incorporated into an annual period: J1 spring beginning (SpB), J7 summer beginning (SuB), J13 autumn beginning (AB), and J19 winter beginning (WB) (Lü 2001). During the following Qin dynasty (221–206 BCE) and Western Han dynasty (206 BCE–8 CE), continuous additions and improvements brought the lunisolar calendar to a total of 24 solar terms (Ge 2014, p. 25). Every term covers 15° of an arc-length along the ecliptic in order to maintain synchronicity with the seasons (Jun 2014, p. 57).

Table 3.7 presents the correspondences between the 24 solar terms and main calendrical elements, such as the 12 terrestrial branches, 12 lunisolar months, and the Sun's 24 ecliptic angles. These solar terms represent different components of the nodal-medial *qi*. They are divided into two groups: 12 major nodal *qi* and 12 minor medial *qi* (He 1998, pp. 213–18; Nielsen 2003, p. 75). The nodal *qi* includes the odd-numbered 12 solar terms, and the medial *qi* includes the even-numbered 12 solar terms. Specifically, every full lunisolar year contains 12 months and 24 solar terms; every month has 2 solar terms (a nodal *qi* and a medial *qi*). In addition, every solar term is divided into three smaller periods, called *hou*, the phenological pentad of a five-day interval (Wu, Yuan). There are 6 hou in a month, and 72 hou in a year. What is more, the first day of every nodal *qi* is the beginning of a month, while the last day of every medial power is the end of the month. Specifically, the first day of every 12 nodal *qi* and every 12 medial *qi* begins roughly the same days, respectively, year after year in the Gregorian solar calendar: the 6th and 21st day of a Gregorian month in the first half, and the 8th and 23rd day in the latter half, with an error of two or three Gregorian days.

TABLE 3.7. CORRELATIONS OF SOLAR TERMS AND CALENDRICAL ELEMENTS

No.	Season	Terrestrial Branch	Month	Ecliptic	Solar-Term *Qi*		
1	Spring	B3	January	315°	J1-SpB	Spring Beginning	Nodal
2				330°	J2-RW	Rainwater	Medial
3		B4	February	345°	J3-IW	Insect-Waking	Nodal
4				360°/0°	J4-SE	Spring Equinox	Medial
5		B5	March	15°	J5-PB	Pure Brightness	Nodal
6				30°	J6-GR	Grain Rain	Medial
7	Summer	B6	April	45°	J7-SuB	Summer Beginning	Nodal
8				60°	J8-LF	Less Fullness	Medial
9		B7	May	75°	J9-GE	Grain in Ear	Nodal
10				90°	J10-SS	Summer Solstice	Medial
11		B8	June	105°	J11-LH	Less Heat	Nodal
12				120°	J12-GH	Great Heat	Medial

TABLE 3.7. CORRELATIONS OF SOLAR TERMS AND CALENDRICAL ELEMENTS (*cont.*)

No.	Season	Terrestrial Branch	Month	Ecliptic	Solar-Term *Qi*		
13	Autumn	B9	July	135°	J13-AB	Autumn Beginning	Nodal
14				150°	J14-HE	Heat End	Medial
15		B10	August	165°	J15-WD	White Dew	Nodal
16				180°	J16-AE	Autumn Equinox	Medial
17		B11	September	195°	J17-CD	Cold Dew	Nodal
18				210°	J18-FD	Frost-Descending	Medial
19	Winter	B12	October	225°	J19-WB	Winter Beginning	Nodal
20				240°	J20-LS	Less Snow	Medial
21		B1	November	255°	J21-GS	Great Snow	Nodal
22				270°	J22-WS	Winter Solstice	Medial
23		B2	December	285°	J23-LC	Less Cold	Nodal
24				300°	J24-GC	Great Cold	Medial

By contrast, figure 3.6 (p. 65) shows that the lunisolar cycle starts from the terrestrial branch Bi-*zi* (hexagram 32 ䷗). For precise identification, all the positive lines (—) of the hexagrams in the figure are drawn in red. Accordingly, the first lunisolar month is November, abbreviated as Nov in the figure. This is the Gregorian month of December, abbreviated as DEC in the figure. The first solar term (not the arithmetic J1) appears this month. It corresponds to J22, winter solstice (WS), with the solar longitude of 270° along the ecliptic. Starting from this lunisolar Nov (equivalently, the Gregorian DEC) of a year, yangqi increases in the first six months and culminates in the lunisolar Apr (equivalently, the Gregorian MAY). This process is seen by those hexagrams with the increasing number of positive lines (—). The increase goes from one monogram to a full set of six in the six hexagrams of 32 ䷗, 48 ䷒, 56 ䷊, 60 ䷡, 62 ䷪, and 63 ䷀.

This waxing (*yang*-ascending) process can alternatively be viewed as a *yin*-descending process. This process illustrates that the northern part of the Earth increasingly receives yangqi, that is, the solar radiation. By doing so, it gradually drives away yinqi until it reaches the maximum.

By contrast, *yin* increases month after month, starting from the lunisolar May (the Gregorian JUN), for the left six months of the year. It culminates in the lunisolar Oct (the Gregorian NOV). This process is represented by the increasing number of the negative lines (--) from one to a full set of six in the six hexagrams of 31 ䷫, 15 ䷠, 7 ䷋, 3 ䷓, 1 ䷖, and 0 ䷁. This waning (*yin*-ascending) process can alternatively be viewed as a *yang*-descending process. It depicts the northern hemisphere of the Earth where yangqi decreases until its minimum.

Mesocosmic World-Ordering Cycle

Concerning the mesocosmic scale, Shao Yong created the most impressive onto-cosmological map in human history. He designed a square-circular diagram, shown in figure 3.3 (p. 57). The figure incorporated the I Ching's numero-cosmology into the lunisolar calendrical system. As discussed earlier, he relied on the numerological progressions of the hexagrammatic *yin* and *yang* to establish a world-ordering system of as long as 129,600 years.

Such a system has a cycle in an order of a normal ice-age period across the earth's surface. It features a set of two correlated tiers, an upper one and a lower one, each of which has four periodicities. The tiers have alternating numerical categories of 12 and 30, comprised of time increments in descending order of duration, grouped in a set of four periodicities (Huang, 1999, pp. 43–44; Deng Qing, p. 15; Wyatt 2010, p. 23) The results were recorded in details in his Treatise of Supreme World-Ordering Principles (as an example, see that by Shao Yong Song-a).

According to the treatise, the upper tier depicts the mesocosmic cycle of Cycle (yuan)–Epoch (hui)–Revolution (yun)–Generation (shi) with:

1 Cycle = 12 Epochs = 360 Revolutions = 4,320 Generations = 129,600 years

which gives us

1 Cycle = 129,600 years;
1 Epoch = 10,800 years;
1 Revolution = 360 years; and
1 Generation = 30 years

The other tier, the lower one, describes the microcosmic lunisolar cycle of year–month–ri–shichen with

1 year = 12 months = 360 *ri* = 4,320 *shichen* = 129,600 *fen*

where

6 ri = 6.0875 days and one fen* = 4 minutes

In either one of the tiers, the alternating 12 and 30 numerical categories are given by the unchanged relations among the four periodicities:

TABLE 3.8. RELATIONSHIPS AMONG THE FOUR PERIODICITIES

	Upper tier	Lower tier	
Cycle	12 hui	year	12 months
Epoch	30 yun	month	30 ri
Revolution	12 shi	ri	12 shichen
Generation	30 years	shichen	30 fen

Scale-Free Cycle

The two categories indicate not only a holographic property which entails the encoding of information from a higher tier precisely onto

*Currently, a fen is used as one minute in Asian countries.

a lower one, but also a holofractal property which suggests a kind of correlative self-similarity across the tiers of different scales (Lefferts 2014). That is to say, entities of nature may be so well self-organized that they follow identical rules and play roles at all times and in all places. Indeed, Shao Yong (e.g., Song-a) demonstrated the holofractal and holographic properties through an ordered scale-free cycle of hexagrammatic *qi*. Table 3.9 lists all the different periodicities in such a cycle, which is identical to the two tiers under discussion.

TABLE 3.9. IDENTICAL SCALE-FREE CYCLE WITH PERIODICITIES OF TWO TIERS

Cycle Year	Epoch Month		Revolution Ri		Generation Shichen		Year Fen		Calendrical Year	
Season	Terrestrial Branch	Sovereign Hexagram	From	To	From	To	From	To	From	To
Winter	B1	H1 ䷗	1	30	1	360	1	10,800	67017 BCE	56218 BCE
	B2	H2 ䷒	31	60	361	720	10,801	21,600	56217 BCE	45418 BCE
Spring	B3	H3 ䷊	61	90	721	1,080	21,601	32,400	45417 BCE	34618 BCE
	B4	H4 ䷡	91	120	1,080	1,440	32,401	43,200	34617 BCE	23818 BCE
	B5	H5 ䷪	121	150	1,441	1,800	43,201	54,000	23817 BCE	13018 BCE
Summer	B6	H6 ䷀	151	180	1,801	2,160	54,001	64,800	13017 BCE	2218 BCE
	B7	H7 ䷫	181	210	2,161	2,520	64,801	75,600	2217 BCE	8583
	B8	H8 ䷠	211	240	2,521	2,880	75,601	86,400	8584	19383
Autumn	B9	H9 ䷋	241	270	2,881	3,240	86,401	97,200	19384	30183
	B10	H10 ䷓	271	300	3,241	3,600	97,201	108,000	30184	40983
	B11	H11 ䷖	301	330	3,601	3,960	108,001	118,800	40984	51783
Winter	B12	H12 ䷁	331	360	3,961	4,320	118,801	129,600	51784	62583

As given by the yellow-shaded cells in the table, the present cycle of the Earth (upper-tier) starts from the Gregorian year of 67017 BCE and ends at 62583. There are four outstanding events in this Cycle:

Revolution 76 was the time of things to start (*kaiwu*);
Revolution 180 was the time when Emperor Yao became the first Chinese ruler;
Revolution 181 was the time when the Xia dynasty began; and
Revolution 315 will be the time of things to terminate (*biwu*).

According to Laozi (chapter 16), free from scales, every cycle can be

> brought to an ultimate at which an unwavering stillness is kept. From this, all things come into being, and go through similar processes of evolution. Then, . . . they respectively return to the same root. The root-returning is called tranquilization, that is, the return to destiny. The fulfillment of the life-resumption is the unchanging law of regularity, the consciousness of which is called being enlightened (my translation).

In such a scale-free cycle, there are five layers in the calendrical distributions, L1–L5, as given in table 3.10.

Layer L1 contains three realms:

- the present realm that dominates an ongoing cycle,
- the past realm that was the transition period from the prior cycle to this ongoing cycle, and
- the future realm that will be the transition period from this present cycle to the subsequent one.

Layer L2 includes four seasons. Borrowing the lunisolar nomenclature, we name the seasons as spring, summer, fall, and winter, which are accompanied, respectively, by the four principal hexagrams: Li 30/45 ䷝, Qian 1/63 ䷀, Kan 29/18 ䷜, and Kun 2/0 ䷁.

Layer L3 consists of six stages. Borrowing the elements related to living beings, the stages can be identified as the following: Embryo (tāi), Genesis (shēng), Development (zhuàng), Deterioration (lǎo), Demolition (sǐ),

TABLE 3.10. CALENDRICAL SCALE-FREE CYCLE

L1 Realm	L2 Season		L3 Stage	L4 Epoch		L5 Solar-Term *Qi*				Ecliptic
				Stem	Branch	No.	Hexagram	Nodal	Medial	
Present	Spring	45	Embryo	S3	*yang* B3	J1	H2	SpB		315°
						J2	H3		RW	330°
					yin B4	J3		IW		345°
		63	Genesis			J4	H4		SE	360°/0°
					yang B5	J5		PB		15°
						J6	H5		GR	30°
	Summer			S4	*yin* B6	J7		SuB		45°
			Development			J8	H6		LF	60°
				S5	*yang* B7	J9		GE		75°
		18				J10	H7		SS	90°
				S6	*yin* B8	J11		LH		105°
			Deterioration			J12	H8		GH	120°
	Autumn				*yang* B9	J13		AB		135°
						J14	H9		HE	150°
				S7	*yin* B10	J15		WD		165°
		0	Demolition			J16	H10		AE	180°
Future				S8	*yang* B11	J17		CD		195°
				S9		J18	H11		FD	210°
	Winter			S10	*yin* B12	J19		WB		225°
			Transformation			J20	H12		LS	240°
Past				S1	*yang* B1	J21		GS		255°
		45				J22	H1		WS	270°
				S2	*yin* B2	J23		LC		285°
			Embryo			J24	H2		GC	300°

and Transformation (huà) (Cai, Song dynasty; my translation). Among the six, the present realm goes from the Genesis stage to the Demolition stage. Before the Genesis stage is the Embryo stage, which interweaves with the past realm. By contrast, the Transformation stage follows the Demolition stage, which interlocks with the future realm.

Layer L4 is ascribed to the 12 epochs in accordance with the 12 terrestrial branches. They are classified into *yang* and *yin* sets. The *yang* set dominates the central parts of all the six stages, including B1-*zi*, B3-*yin*, B5-*chen*, B7-*wu*, B9-*shen*, and B11-*xv*. The *yin* set is dominated by the boundary parts of two adjacent stages, including B2-*chou*, B4-*mao*, B6-*si*, B8-*wei*, B10-*you*, and B12-*hai*. They also obey the chain of the 12 sovereign hexagrams: H1-*fu* 32 ䷗, H2-*lin* 48 ䷒, H3-*tai* 56 ䷊, H4-*dazhuang* 60 ䷡, H5-*guai* 62 ䷪, H6-*qian* 63 ䷀, H7-*gou* 31 ䷫, H8-*dun* 15 ䷠, H9-*pi* 7 ䷋, H10-*guan* 3 ䷓, H11-*bo* 1 ䷖, and, H12-*kun* 0 ䷁.

The last layer, L5, gives the 24 world-ordering terms corresponding to the 12 terrestrial branches. The names of these terms borrow those of the lunisolar terms distributed respectively among 24 solar longitudes along the ecliptic. Among them, every medial *qi* and the following nodal *qi* belong to a sovereign hexagram. The whole set of the 12 terrestrial branches differentiate the 24 terms from each other through describing the series of stages interlacing with the set of 10 celestial stems (Drasny 2011, pp. 39–70; He 1998, p. 262; Xu 1997, pp. 199–220; Yu, 2004, pp. 610–19):

1. The terrestrial branch B3-*yin* refers to the stage of initiation. This stage interlaces with the celestial stem S3-*bing* to indicate a vigorous growth. It lasts two terms: J1 Spring Beginning (SpB) and J2 Rain Water (RW), during which *qi* increases rapidly.
2. The terrestrial branch B4-*mao* refers to the stage of replenishment. This stage does not interlace with any celestial stems. It lasts two terms: J3 Insect Waking (IW) and J4 Spring Equinox (SE), during which *qi* keeps growing.
3. The terrestrial branch B5-*chen* refers to the stage of accomplishment upon replenishment. This stage does not interlace with any

celestial stems. It lasts two terms: J5 Pure Brightness (PB) and J6 Grain Rain (GR), during which *qi* becomes strong.

4. The terrestrial branch B6-*si* refers to the stage of removal of any over-replenishment. This stage interlaces with the celestial stem S4-*ding* to indicate maintaining substantial growth. It lasts two terms: J7 Summer Beginning (SuB) and J8 Less Fullness (LF), during which *qi* drives the community of organisms to form and develop in the local heaven-earth cosmos.
5. The terrestrial branch B7-*wu* refers to the stage of stabilization. This stage interlaces with the celestial stem S5-*wu* to indicate a luxuriant ripening growth. It lasts two terms: J9 Grain in Ear (GE) and J10 Summer Solstice (SS), during which *qi* begins to exhaust and decline after reaching its climax.
6. The terrestrial branch B8-*wei* refers to the stage of controlling. This stage interlaces with the celestial stem S6-*ji* to indicate the onset of decay. It lasts two terms: J11 Less Heat (LH) and J12 Great Heat (GH), during which *qi* begins to fade away.
7. The terrestrial branch B9-*shen* refers to the stage of precariousness. This stage does not interlace with any celestial stems. It lasts two terms: J13 Autumn Beginning (AB) and J14 Heat End (HE), during which *qi* and the community of organisms begin to destruct.
8. The terrestrial branch B10-*you* refers to the stage of collapse. This stage interlaces with the celestial stem S7-*geng* to indicate the withering to decline and fall, opposite to S2-*yi*. It lasts two terms: J15 White Dew (WD) and J 16 Autumn Equinox (AE), during which *qi* and all organisms begin to decline and break down.
9. The terrestrial branch B11-*xv* refers to the stage of extinction. This stage interlaces with the celestial stem S8-*xin* to indicate the preparation for a new turn. It lasts one term only: J17 Cold Dew (CD), during which everything disintegrates without exception.
10. The prior stage of the terrestrial branch B11-*xv* interlaces with the celestial stem S9-*ren* to prepare the hatch of *qi*. It lasts

one term: J18 Frost Descending (FD), during which the local heaven is completely extinct.

11. The terrestrial branch B12-*hai* refers to the stage of completion. This stage interlaces with the celestial stem S10-*gui* to indicate the awakening of *qi*. It lasts two terms: J19 Winter Beginning (WB) and J20 Less Snow (LS), during which everything, including the local earth, begins to ruin into a void where only the primordial *qi* exists.
12. The terrestrial branch B1-*zi* refers to the stage of origination. This stage interlaces with the celestial stem S1-*jia* to indicate the continuous origination of *qi* from the void. It lasts two terms: J21 Great Snow (GS) and J22 Winter Solstice J22 (WS), during which the scale-dependent local celestial realm initiates its formation.
13. The terrestrial branch B2-*chou* refers to the stage of preterm. This stage interlaces with the celestial stem S2-*yi* to indicate a weak growth of everything out of *qi*. It lasts two terms: J23 Less Cold (LC) and J24 Great Cold (GC), during which the scale-dependent local terrestrial environment commences its construction.

Generalized Pattern of Scale-Free Evolution

The scale-free cycle shown in table 3.10 illustrates how the entities of natural things at any scale follow the same pattern of Birth (shēng) in the scale-free spring, Growth (zhǎng) in the scale-free summer, Harvest (shōu) in the scale-free fall, and Storage (cáng) in the scale-free winter once evolving within the three realms, past, present, and future. This pattern can be classified into the above mentioned six specific stages, i.e., Embryo, Genesis, Development, Deterioration, Demolition, and Transformation. It gives an insight into the natural wholeness of the evolving reality (Cheng 2019; Von Humboldt 1866, pp. 106–18). This wholeness is expressed by a generalized scale-free cycle that offers a holographic and holofractal evolution irrespective of the sizes of any entities concerned.

Such a scale-free cycle of the evolving pattern is illustrated in figure 3.7 (c.f. Xu 1997, p. 198). The pattern fits with any entities (such as a person, a society, a galaxy). Red arrows in the figure indicate the evolving direction. The figure discloses that the life of any entity proceeds in a similar cycle from voidness to voidness by experiencing expansion and contraction:

- an entity first emerges from the resulting void of a prior cycle;
- it then grows and expands to the maximum;
- it afterward experiences a contracting and decaying process;
- until it finally returns to the void again for the beginning of the next cycle.

Such a cycle keeps going repeatedly from the past infinity to the future infinity. The cyclic evolution never stops coming from the void and returning to the void. According to Shao Yong's treatise, we conclude that every cycle owns 129,600 scale-free units that are in accordance with 60 ruling hexagrams in periodicities falling within 12 and 30 categories.

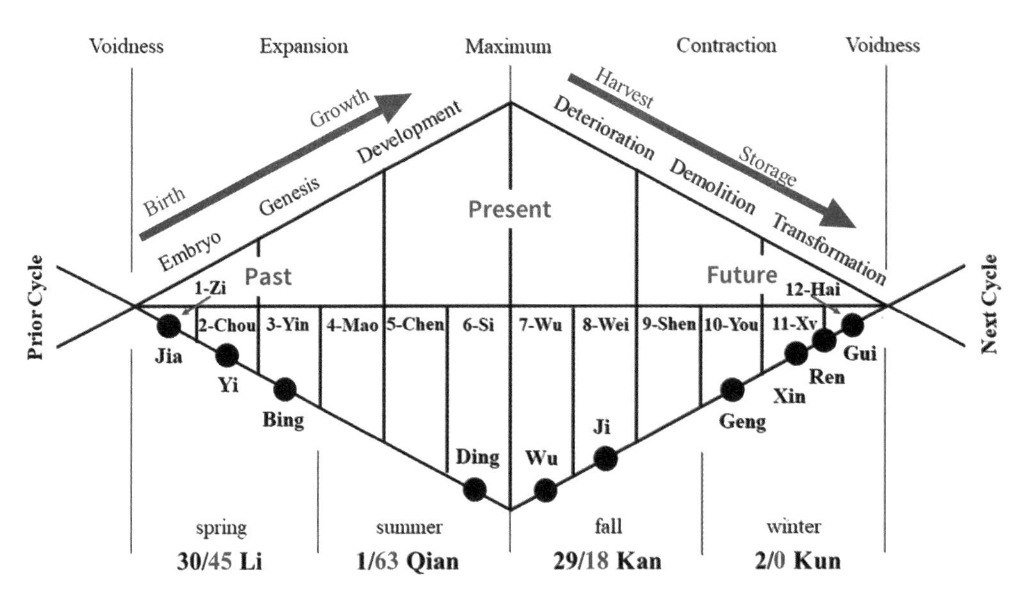

Figure 3.7. Generalized pattern of evolution in a scale-free cycle

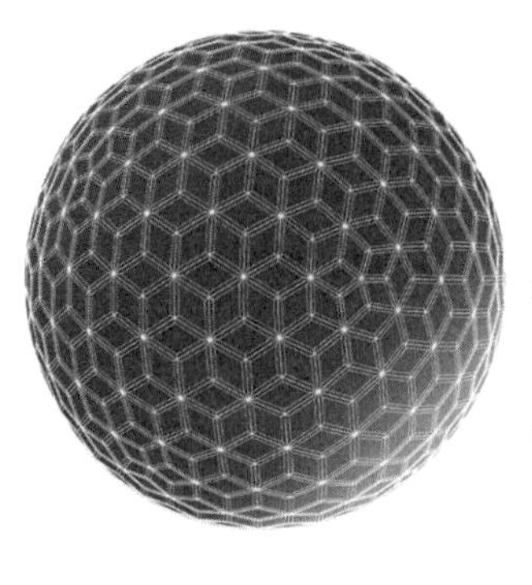

4 Swimmean Spiritual Evolutionism

EASTERNIZATION OF CHRISTIANITY

After the catastrophic Great Famine (1315–1317) and the devastating Black Death (1347–1351), Europe entered the period known as the Renaissance, which lasted from the fourteenth to the seventeenth centuries. The Renaissance witnessed flourishing in philosophy, arts, culture, and science, and marked a transformation from an inwardly spiritual, decentralized, and hierarchical world to an outwardly secular, centralized, and individualistic one (Tuchtey 2011). In particular, the colonial Age of Discovery began from the fifteenth century in America, Africa, and Asia. During that time, Martin Luther advocated a revolutionary Protestant theology (Chambers et al. 2007).

During the seventeenth century, Protestant Christianity emphasized that all beings are the fruits of salvation with the ultimate universal reconciliation in God, and that they will attain the final holiness and happiness from the grace of God through the ministry of his Son, Jesus Christ (Sawyer 1882, p. 15; Strohman 2017, pp. 143–44). As a result, theistic universalism emerged and the spread of Christianization moved eastward, leading to historically significant Sino-Western contacts through Christian missions. Yet it's important to note that easternization had already occurred much earlier.

The first traceable evangelization appeared in 635 during the Tang dynasty (618–907). The Eastern Roman Empire dispatched a team of missionaries, called the Nestorians, to China (Mikkelsen 2006). This earliest Sino-Western contact endured for 150 years (Zhu 1993, pp. 79–97).

The second contact came after the split of the Roman and Eastern churches in 1054. The split happened owing to the East-West Schism in Rome. A Catholic missionary, John of Montecorvino, arrived in Beijing in 1294 during the Yuan dynasty (1271–1368). Serving as the archbishop, he stayed there for several years and established some of the earliest Catholic churches in the region (Barnard and Hodges 1958, pp. 107–8).

The third contact began in 1552 during the Ming dynasty (1368–1644). Among the missionaries, an Italian, Matteo Ricci, came to China in 1582. He extensively borrowed Daoist, Buddhist, and Confucian words and terms to explain Christian doctrines and Western astronomy. He also widely quoted Chinese classics to emphasize the ontologically supreme deity of Christianity. More impressively, he tactically employed the historical accomplishments that best represented ancient Chinese civilization to symbolize those of Western civilization, highlighting a connection and commonality by framing both as manifestations of the Lord's achievements (Spence 1985). His plan for universal salvation in China was laid out in his *Complete Corpus* in Chinese (Zhu, 2007) and *The True Meaning of the Lord of Heaven* (Ricci 1985). With such attempts at tapping into the local culture, he expected that the Christian God could easily be understood and accepted by the Chinese people, and thus, that God could eventually wash off the traditional deities and beliefs in Chinese culture.

However, the power of Christianization was too weak to shake the traditional beliefs, customs, and thoughts that had survived in China for thousands of years. For more than three hundred years after Matteo Ricci, Christian missions failed to have any significant impact on native religious thinking. Similar to Matteo Ricci, missionaries coming from the West encountered the abstract Chinese atheistic philosophy, yet they saw it garbed in polytheistic religious sects and rituals. It was difficult for them to understand the mysterious Eastern culture and its various indigenous customs. Facing such an embarrassing situation, more and more missionaries increasingly shifted their stand from Christianizing the East by replacing indigenous cultures to developing a so-called religious syncretism, that is, amalgamating Chinese culture under the umbrella of Christianity. Dominated by this innovative ideology of evangelization, the fourth wave of Sino-Western contacts came into play after 1840. The era followed the European transition to machine-based manufacturing and military industries and the expansion in trade and transportation, developments in agriculture, and scientific discoveries (Falkus 1987, pp. 1–12).

During the seventeenth to nineteenth centuries, some missionaries who had learned the Chinese language and absorbed Chinese culture began to spread Chinese classics back to the West (Hegel 2011, pp. 212–13, footnote 2; p. 218, footnote 24; p. 219, footnote 26). Among these classics, the I Ching was first translated into Latin in 1626 as one of Confucius's Five Classics (Vieira 2001, p. 223; Zheng and Yue 2011). In 1658, the first diagram of the 64 hexagrams appeared in Europe (Martini 1658). In 1687, the translated version of the divinatory interpretations of the 64 hexagrams was published there (Redmond and Hon 2014, p. 194).

By comparison, Laozi's Tao Te Ching was first translated into Latin in 1788 by Jesuit missionaries in China. The masterpiece was believed not only to pose "a famous puzzle which everyone would like to feel he had solved" (Welch 1965, p. 7), but also to disclose that "the mysteries of the most holy trinity and of the incarnate God were anciently known to the Chinese nation" (Hardy 1998, p. 165).

In 1814, the first chair of Sinology appeared at the Collège de France. The chairman was a French Sinologist, Jean-Pierre Abel-Remusat. He studied Laozi and translated the Tao Te Ching into French in 1820. He suggested that the Dao indicated the ontological principle that conveyed the triple sense of Supreme Being, reason, and word (Hegel 1892, p. 124; 1983, pp. 122–23, 126–27). More interestingly, he argued that the syllables of three Chinese words, *yí-xī-wēi*, used to express "colorless," "soundless," and "shapeless," come in fact from the Latinized name of the God "*Je-ho-wah*" (Hegel 1892, p. 125).

During the eighteenth to twentieth centuries, the mysterious metaphysical perspective of the I Ching and the Tao Te Ching attracted more Westerners. They applied the ancient Chinese wisdom to push forward Western civilization in science, philosophy, and psychology. A notable example occurred in the work done by Leibniz. In 1703, he applied the square-circular diagram (fig. 3.3) to establish the binary numbering system (Hu and Li 2004, 2006). Another example was in the works of Georg Wilhelm Friedrich Hegel. He made serious references to the I Ching, the Tao Te Ching, as well as additional Chinese works of literature in his own writings, such as his *Phenomenology of Mind* and *Greater and Lesser Logic*, though without any citations (Ma 2016). He eventually acknowledged his references to the Chinese sources in 1822 (Kim 1978). More conspicuously, Carl Jung made use of the I Ching's cosmological presuppositions to describe an acausal connection between events for his principle of synchronicity, in sharp contrast to the traditional notions of cause and effect. The new principle might adequately explain the psychic parallelisms in the psychology of the unconscious (Jung 1947, pp. 142–43).

Though most missionaries were attracted by ancient Eastern culture and contributed to the spreading of Eastern learning to the West, they never lost their faith. They believed that God was the only loving One and that he would not alienate anyone from his presence (Richard 1890). To strengthen the beliefs in Christian universalism among Chinese followers, their endeavors included the idea that all non-Christian religions coexist together in the kingdom of God, and that every faith and every culture has its version reflecting the understanding of His love (Hitchcock 2013, pp. 1–13). This evangelical strategy did intensify more or less the vitality and superiority of Christianity, owing to its adoption and insistence of the declaration of faith that God is the ultimate One for all beings on Earth.

Under the umbrella of this universalism, the critical Daoist concept, *dao*, was thought to be the same as the Western principle of ontology that was suggested by Aristotle's theory of substances—that is, things or bodies—and accidents—that is, qualities, events, or processes (Smith 2001, pp. 79–97; Wang 1997, pp. 437–44). Besides, the Messianic teachings of early Christianity had been shown to successfully permeate and merge with the salvific will of Indian Buddhism. Thus, the prevalence of Buddhism in China was regarded to channel the Christian salvation messages of hope and redemption (Richard 1890).

More intriguingly, it was claimed that the fundamental five Confucian virtues (wǔ cháng)—benevolence (rén), righteousness (yì), propriety (lǐ), wisdom (zhì), and integrity (xìn)—are the gifts offered through God's mercy to the East, yet using non-Christian nomenclature; while, at the same time, God offered the same to the West through the Scriptures (Chao 1989, p. 63). Therefore, though wearing different garbs, or speaking different languages, it was taken for granted that all humans live together in the sacred realm, the Kingdom of Divine Love, and demonstrate common moral and ethical principles across different regions and cultures (Soothill 1924, p. 17, 127–29).

Consequently, all beings were considered together to form a Christianity-leading league on Earth, and the league spread through colonialism across continents and led the secular world by virtue of the peerless Western civilization (Schlesinger 1974, p. 342).

ONTO-PANENTHEISM AND COSMOGENETIC TRINITY

During the campaign for Christian universalism, ancient Western ontotheology gained an unprecedented ally. It integrated Eastern philosophical ideology, especially the Daoist onto-cosmology, to develop a so-called onto-panentheism. Rather than the out-of-date pantheistic notion that all, including the universe itself, is God and God is all, the onto-panentheistic philosophy suggested that all is *in* God, and God is *in* all (Mickey 2016, p. 41).

The word *panentheism* first appeared in 1809 and became popular in 1829 (Clayton 2010; Gregersen 2004, p. 28). However, there had been no clear onto-panentheistic vision until Teilhard de Chardin, a missionary in China (who would gain more renown there as one of the leading finders of the Peking man's skull). He reconciled his idea of the noösphere (the sphere of human thought) with cosmological evolutionism (Fuentes 1996; Levit 2000). He lived in China for twenty-four years from 1923 to 1946, but made frequent trips outside the country. During those years, he finished his major opus entitled *The Phenomenon of Man* (Teilhard 1975a, 1999). All of his letters, journal entries, and books expressed the mystical influence of Eastern religion and culture on his unorthodox interpretation of the cosmic evolution and dynamics of reality (Bidlack 2010). In this reality, all of matter, life, and mind were believed to be the ingredients that demonstrated the sacredness of the divine presence (Birx 1999). These were seen to have emerged collectively from the onto-panentheistic point, God-Omega, a previously mentioned Christian term coined to describe the unity of a hybrid material-spiritual reality (Castillo 2012; Teilhard 1975b, pp. 134–47). The term, though alien in the East, embraces the notion that onto-panentheism intrinsically expresses both the ontological and the cosmological essence of *dao* rooted deeply in the Eastern philosophy.

In the ontological aspect, on the one hand, the ontological principle of the onto-panentheism is equivalent to the Daoist *li* (principle) of taiji (the Great Ultimate), which can be portrayed as nothingness → beingness → nothingness (Lane 1996). First, the universe is a divine milieu, and the milieu is the transcendent body, that is, the culminating convergence of the all that is beyond evolution. Second, this transcendence can be named God-Omega, the point of unity, that is, the oneness that is the end of the universe. Third, point Omega also features the nothingness of a new beginning of the next cycle, Alpha. Finally, at

Alpha, there exists collective humankind to advance to a spiritual union with a personal God in a distant future.

In the cosmological aspect, on the other hand, it was argued that the evolving base is in a process that harmonizes and synthesizes different entities or things; and, this process is first maintained through geogenesis, then biogenesis, and, finally, noögenesis (Luquet 2006). Furthermore, the resultant biosphere surrounds the geosphere to make a planetary and mystical Christogenesis; and the resultant noösphere envelopes both the geosphere and the biosphere to engulf the Earth and detach itself from the globe but immerse itself into the final God-Omega through transcending space and time (Birx 2015; Grim 2016). In these terms, "Christogenesis" refers to the process that the whole universe becomes a cosmic body of Christ (e.g., Udías 2009).

According to Teilhard, onto-panentheism owes its onto-cosmological focus to a universe that is "no longer a state, but a process where humanity is one part of its cosmological products" (Teilhard 1964, p. 261; 1975b, pp. 134–47). This process was regarded as an integrated physical-biological-spiritual evolution, driven by the cosmic energy that consists of components of "mind" and "matter" (Teilhard 1975a, pp. 25, 64; 1975b, pp. 134–47).

Placing classical mechanics (Kepler's laws of planetary motion and Newton's law of universal gravitation) within Christianity, Teilhard treated the "mind" component as a cosmic coherent "tangential energy" to connect all things of the same order as a whole, while regarding the "matter" component as a cosmic attractive "radial energy" to drive a world transforming from lower complexity and centricity to a higher one (Teilhard 1964, p.116; 1999, p.186). See the upper part of figure 4.1.

Influenced enduringly by Teilhard, Thomas Berry was one of the most renowned Asian scholars in the West. He studied Asian thought and religion in China from 1948 to 1949 after his doctoral study in the United States. He was an expert in the spiritual dimensions of Asian classics, especially Confucianism and Buddhism. Compared to Teilhard, Berry went to a deeper level of seeing into "the vision of a unified evolutionary process in which the human participates in the ongoing developmental unfolding" (TBF 2022).

Concerning the "radial energy" articulated in Teilhard's process cosmo-theology, Berry elaborated that the energy can be further divided into two parts in response to the physical and psychic dimensions, respectively. In Teilhard's onto-panentheistic universe, Berry suggested that the two parts were interdependent: they can fold into each other, yet neither is collapsible into the other nor separable from the other (Mickey 2016, p. 41). See the lower part of figure 4.1. It is clear that Berry borrowed

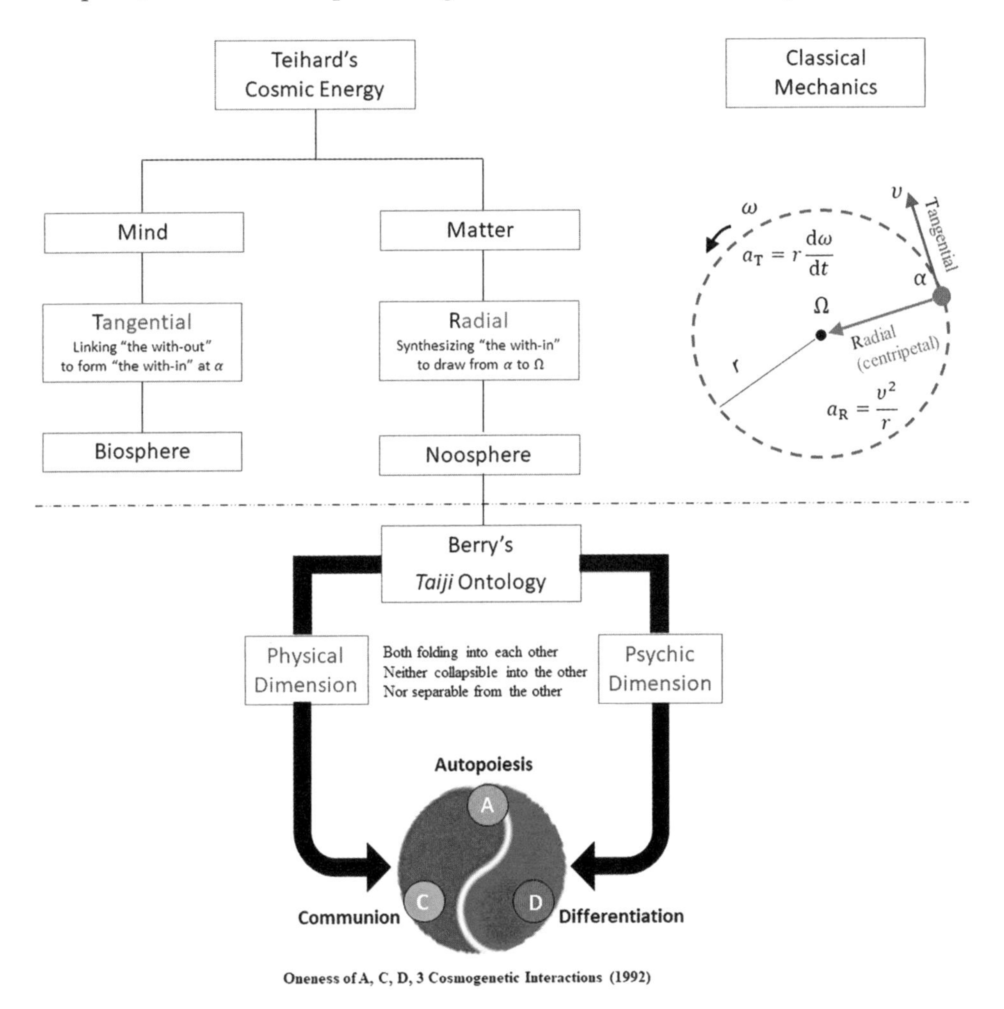

Figure 4.1. Teihard's cosmic energy and Berry's taiji ontology

Taiji's *yin-yang* properties to describe the mutual coupling between the physical part and the psychic part of the "radial energy."

Following Berry, Swimme inherited and developed Teilhard's philosophy that cosmic evolution is a conscious transfiguration. As an evolutionary cosmologist, Swimme integrated science and spirituality to understand the nature of the universe. Together, Berry and Swimme proposed a trinitarian cosmogenetic principle, known as the Berry-Swimme trinity of ontological cosmogenesis. This principle governs the leading theme and the basal intentionality of all existence (Kelly 2015). It deals with a set of three primal cosmic powers resulting from the mutually interacting physical and psychic parts of the "radial energy" (again, see the lower part of figure 4.1 for reference): differentiation (D), autopoiesis (A), and communion (C) (Swimme and Berry 1992, pp. 71–73). Differentiation is equivalent to Homeostasis (or a homogenized uniqueness). The equilibrium it embraces means things are different or autonomous, with variations, disparity, and multiform natures demonstrating heterogeneity and articulation with a complexity that leads to diversity. As entities start to differentiate, they begin to self-organize into recognizable forms, a procedure called autopoiesis. It is equivalent to emergence or self-creativity (or auto-generation, self-production) of differentiated objects, complete with their subjectivities (e.g., structure and property) that make each unique among the others. At the same time, these entities become interrelated with each other and display features of communion, that is, interrelatedness or interdependence.

Thus, the universe orders itself by differentiating itself into different recognizable aspects or parts (i.e., differentiation or uniqueness), each of which not only orders and regulates itself internally (i.e., autopoiesis or self-creativity) but also organizes itself with regard to all its aspects—in other words, communion or interdependence (Midgley 2007, p. 42). This trinity prevents the universe from becoming a "homogeneous smudge" via the power of differentiation; what is more, by means of "an inert and dead extension" via the power of autopoiesis, it differentiates itself from a series of "isolated singularities of being" via the power of communion (Mickey et al. 2017, p. 150).

SWIMMEAN COSMOGENETIC POWERS

Swimme extended his exploration into the centuries-long investigations regarding the interconnection of cosmological evolution and human existence. He drew upon insights from ancient religious understandings and modern scientific advancements in philosophy, cosmology, and consciousness. Through a thorough analysis and synthesis of significant contributions by others, he revealed that the dynamic interplay among three fundamental powers—D, A, C—results in nine additional elements: Seamlessness, Centration, Allurement, Synergy, Transmutation, Transformation, Cataclysm, Radiance, and Collapse, all of which were considered as the keys to understanding how the microcosm mirrors the macrocosm, for example, from human DNA to star systems, or from bodily lungs to galaxies (Center for the Story of the Universe 2004), not to mention their uniqueness in experiencing "a spiritual path or progression, leading to a fundamental appreciation for the beauty" of the universe (Slocum 2006, p. 12).

These powers were elaborately examined in several publications (e.g., Swimme 2017b; Swimme and Anderson 2004; Amberg 2011, pp. 7–39; Le Grice 2011, pp. 22–23; Ma 2018b). They lent prominence not only to integrating ancient pearls of wisdom with modern disciplines to address potential cosmo-ecological challenges (Grim and Tucker 2017, pp. 1–14), but also, due to their enduring presence, to the universe's engagement in human-ecological evolution through the intertwining of science with the humanities (Bod and Kursell 2015; Swimme and Tucker 2011, pp. 71–73).

The definitions of this set of powers were clearly articulated in a 3-DVD set of lectures (Swimme and Anderson 2004). These lectures presented a positive and life-affirming vision of human potential, along with a heightened sense of cosmo-ecological responsibility and interconnectedness, while the powers are believed to be responsible for the perpetual cyclic evolution of the universe, spanning from the beginning of time and encompassing all forms of life. Below is a detailed description of these powers.

- **P1, Seamlessness.** This is the power of all the other powers and the generative mainstay of reality. It is the all-nourishing abyss of pure nothingness that is ultimate or boundless before the revitalization and rebirth of a universe through an explosion of elementary particles in space, followed by the formation of primary elements found in all things. It is the quantum vacuum that has the potential energy to support nonlocal connections or quantum entanglement (DVD-episode 1).
- **P2, Centration.** This is the self-centering power of individual entities or centers to maintain or nurture the development of life for the birth of individuated beings. It is the ability to form new centers of wholeness, making their manifestation. It is shown in homologous ways at different levels and orders of magnitude in the universe (DVD-episode 2).
- **P3, Allurement.** This is the attraction power that allows real entities to hold together and be drawn into a relationship of intimacy and union. It is the primordial attractive force (e.g., gravitation) in the cosmos that pervades all life, matter, space, and time, and holds entities together. It is the energy that enables one to come into a deeper relationship with the cosmos (DVD-episode 3).
- **P4, Emergence.** This is the creative power allowing real entities to transcend themselves beyond the established limits. It is the capacity to bring forth new forms and qualities at all levels, featuring unique properties of complexity owing to continually arising self-assembly and self-organization (DVD-episode 4).
- **P5, Synergy.** This is the cooperation power for real entities to enhance mutual relationships in larger wholes, arrangements, or contexts and thus acquire more free energy, creativity, and productivity. It strengthens the vitality of things and leads to the

deepening of diversity, complexity, and resilience in all living systems in collaborative relationships with others (DVD-episode 7).

- **P6, Transmutation.** This is the adaptive power that allows individual entities to accommodate the community. It is the energy of the universe that forms or shapes micro- or macro-ecosystems in response to the requirements of the whole community (i.e., natural selection). It is the fundamental process of individuation that encourages individuals to pursue a more profound incarnation into an expanded context (DVD-episode 8).
- **P7, Transformation.** This is the development power that allows for individual entities to upgrade the whole. It is the drive of an individual entity to activate communion, community, and intimacy collectively, marked by an enhanced coherence with other powers and an expanded potential for coevolution (DVD-episode 9).
- **P8, Cataclysm.** This is the destructive power inherent in real entities to do away with obsolete properties while strengthening the capacity of things to work together to create new forms and encourage the proliferation of interconnected relationships. However, the fluctuations within an entity can become too great to maintain the existing system (DVD-episode 6).
- **P9, Homeostasis.** This is the maintenance power for real entities to keep uniquely differentiated properties in their evolution. It is the energy or the balancing force that serves to conserve the particularity and vitality of things in forms and processes that are the most valued through trial and error over millions of years, making them resonant with the surroundings at all levels of consciousness (DVD-episode 5).
- **P10, Interrelatedness.** This is the care or humility power for entities to nurture or take care of relationships with others that leads to a deepening of bonds and life. It follows the anthropic principle to evoke new directions. It is the instinct (or decision) to nurture a process that leads to increased intimacy, resonance, communion, differentiation, and complexity (DVD-episode 10).

- **P11, Radiance.** This is the magnificent power that entities have to communicate in the deepened Interrelateness. It is the energy received by others at both a surface level and in the depths of perception. It is the experience of resonance and reverberation with natural phenomena. It is the celebratory revelation of the deep mystery of the cosmos. And it is the conscious revelation of the termination of reality (DVD-episode 11).
- **P12, Collapse.** This is the annihilation power of reality leading it to undergo termination until a state where all the entities fall apart, all the forms are broken down, and reality itself decomposes to a "pure" nothingness where no reality exists; that is to say, no matter, no energy, no laws, no space, and no time. Seamlessness may emerge from it depending on the quantum probability (DVD-episode 11).

These twelve cosmogenetic powers apply to all entities. They possess several important features such as closed interdependence, pair reflexivity, pair complementarity, and cause-effect relations.

Closed Interdependence

The dynamic relationship among the powers is effectively underscored by a closed chain of interdependence, as illustrated by the dark-red arrows connecting the powers in figure 4.2. Each power relies on the preceding one, forming an intricately linked sequence that contributes to and influences the manifestation of the next in a cohesive and interconnected manner.

The chain gives a cyclic sequence as follows: Seamlessness → Centration → Allurement → Emergence → Synergy → Transmutation → Transformation → Cataclysm → Homeostasis → Interrelatedness → Radiance → Collapse → Seamlessness (new cycle). In one cycle, every power is the effect of the prior one; at the same time, it behaves as the cause of the following one.

However, accurately speaking, we should keep in mind that this

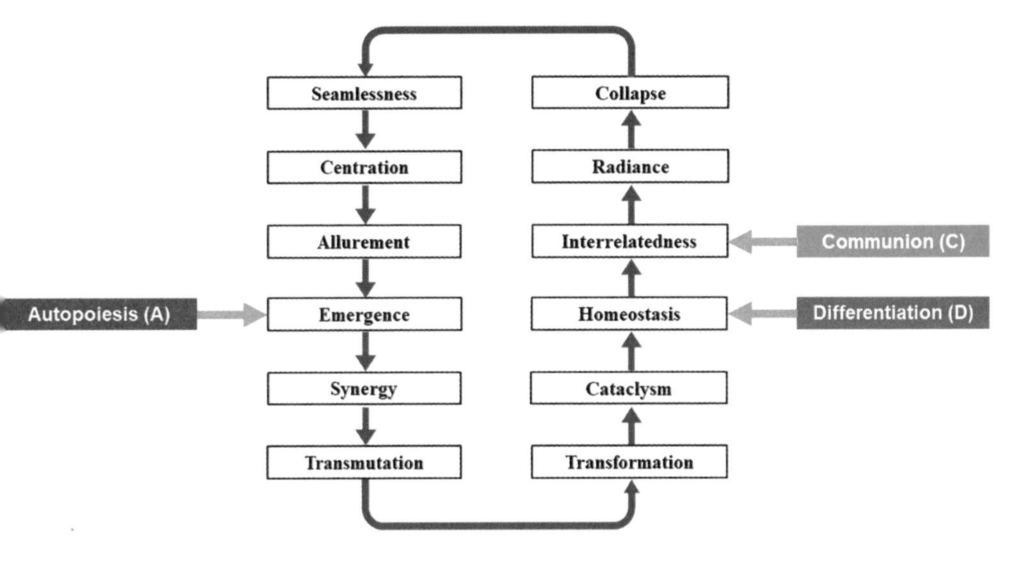

Figure 4.2. Closed interdependence of Swimmean powers

interdependence pattern does not always describe all the evolving processes of entities. In some biological circles—for example, in the caterpillar-butterfly growth—a few powers may be too weak to be unfolded.

Pair Reflexivity

The twelve powers also form six reflexive pairs, exerting influence on entities. These pairs are indicated by the six blue arrows in figure 4.3 (p. 92) and listed as follows:

1. Seamlessness from pure nothingness ↔ Collapse to pure nothingness
2. Centration toward wholeness ↔ Radiance against wholeness
3. Allurement from primordial causality ↔ Interrelatedness for holistic causality
4. Emergence from self-assembly and self-organization ↔ Homeostasis of individuation to overcome self-imposed limitations
5. Synergy for construction ↔ Cataclysm for destruction

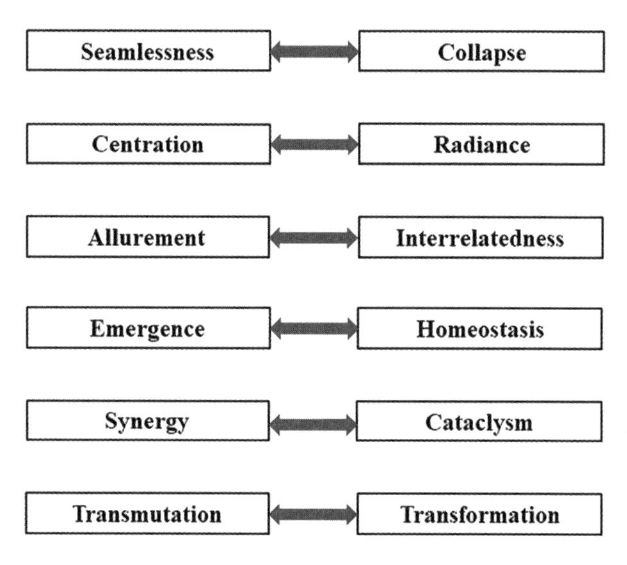

Figure 4.3. Pair reflexivity of Swimmean powers

6. Transmutation to accommodate community ↔ Transformation to dominate the community

We point out here that the names of the twelve powers are somewhat fluid rather than fixed in describing the above pair-reflexive relations. The innate overlapping and mutually implicated powering processes may have happened, may be happening, or will happen at different levels and scales. Notwithstanding this fact, the terminological definitions are still distinct enough to demonstrate the subtleties and differentiate the respective primal actions of all the powers (Le Grice 2011, p. 224).

Pair Complementarity (Mirroring)

In driving the evolution of entities, the twelve powers contain six pairs which complement each other rather than mutually exclude each other. Refer to figure 4.4, where Birth, Growth, Harvest, and Storage—the four segments of the scale-free cycle given in figure 3.7 (p. 77)—are labelled on both sides of the powers. On the left-hand side, a group of

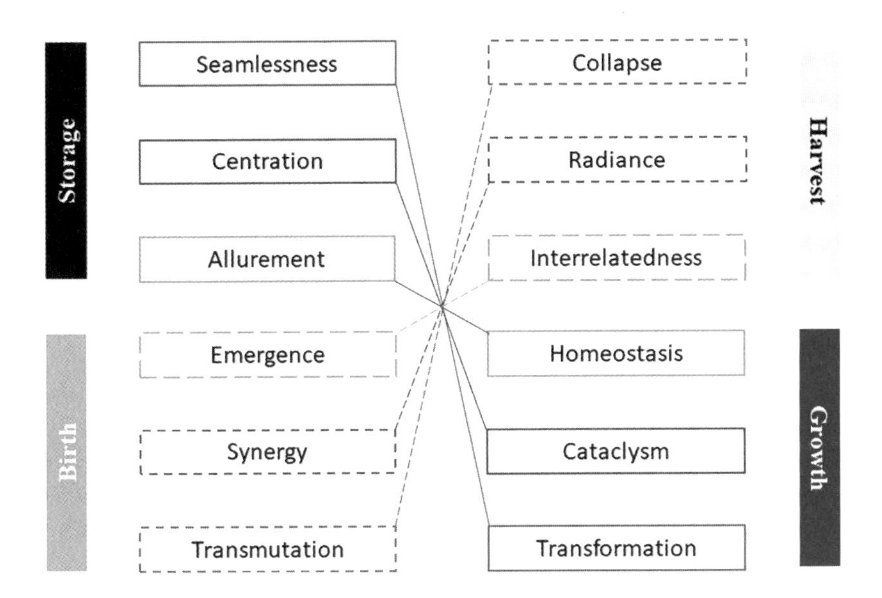

Figure 4.4. Pair complementarity (mirroring) of Swimmean powers

six powers propels a process from nothingness to beingness, while on the right-hand side, another group of six powers facilitates an opposite process from beingness to nothingness. Considering figure 4.2, where all the powers form a cascading procedure to drive the evolution of things, the synthesized power of the following opposing pairs will certainly be complementary, that is, remain invariant:

1. Seamlessness from pure nothingness—Transformation to dominate the community
2. Centration toward wholeness—Cataclysm for destruction
3. Allurement from primordial causality—Homeostasis of individuation to overcome self-imposed limitations
4. Emergence from self-assembly and self-organization—Interrelatedness for holistic causality
5. Synergy for construction—Radiance against wholeness
6. Transmutation to accommodate community—Collapse to pure nothingness

Cause-Effect Relation

Pervading the three realms (past, present, and future) of all evolving things, the twelve powers are characterized by dual causality* in alignment with their connections not only to these realms, but also to the three primal sources (D, A, C), as well as the three components in the evolutionary process (input, action, output). Refer to figure 4.5 for more details. The dual causality includes:

- Cause 1 in the past, consisting of P1 and P2;
- Effect 1 in the present, consisting of P3, P4, P5, P6;
- Cause 2 in the present, consisting of P7 and P8;
- Effect 2 in the future, consisting of P9, P10, P11, P12.

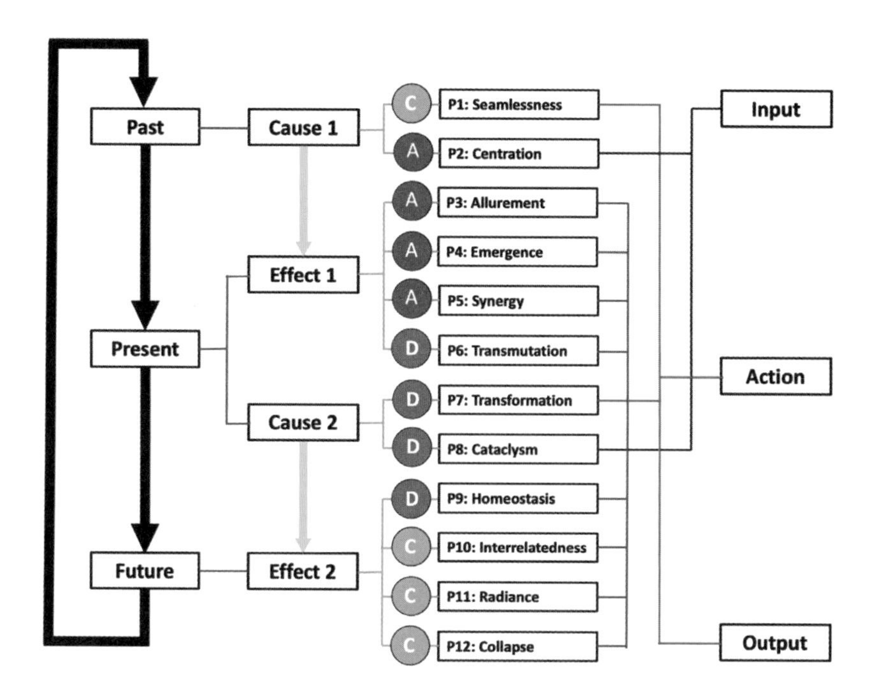

Figure 4.5. Cause-effect relation of Swimmean powers. Recall that D, A, and C stand for differentiation, autopoiesis, and communion.

*For an in-depth discussion of the cause-effect relation, refer to my next companion book, which I have currently titled *The Tao of Consciousness: Quantum Principles of Reincarnation*.

QUANTIFICATION OF THE TWELVE SWIMMEAN POWERS

The European Space Agency (ESA) ran the Planck space observatory from 2009 to 2013. The mission proved that the observable universe is composed of ordinary matter (4.8 percent), neutrinos (0.1 percent), cold dark matter (26.8 percent), and dark energy (68.3 percent) (Lochner 1998; Planck Collaboration 2014), while the critical density (ρ_{crit}) out of measurements is 8.5×10^{-27} kg/m^3 (Planck Collaboration 2016). Consequently, the calculated total mass-energy (ε) of the universe is $\approx 10^{69}$ J, equivalent to a total mass (M) of approximately 26 to 65 billion Milky Way galaxies if their masses are assumed the same as that of our galaxy, 2.0×10^{42} kg.

The twelve powers quantify ε with a series of twelve epochal powers, known as *quanlia* (*quanle** in the singular form noted as Q hereafter) in the unit of Watt (W). If an epoch lasts ΔT (second) in time, we have $Q \cong \Delta Q = \varepsilon / \Delta T$. Writing τ = 129,600, then:

$$\Lambda_S = \frac{1}{(540\tau^8)} = 2.33 \times 10^{-44} \text{ (second)}$$

which yields the Q-magnitudes of the twelve epochal powers.

The Q-magnitudes of the twelve epochal powers are given in table 4.1 (p. 96), where the related corresponding cosmological epochs of a big-bang eon are given as a reference, respectively.

EVOLUTIONARY COSMOLOGY

The set of Swimmean powers offers a scheme of universal cosmological evolution of conscious nature. It elucidates a twelve-stage pattern of cosmo-ecological evolution, driven by respective cosmogenetic powers,

*By contrast, *quale* (*qualia* in plural form) is used in cognitive science to describe subjective, conscious experiences; note that quale is dimension-free, without a physical unit.

for all realities at different scales of existence, such as the human bodily system, the solar-terrestrial system, the galactic system, and so forth. It fosters an innovative dialogue between science and the humanities in terms of a variety of known natural laws or principles in cosmology, physics, anthropology, and beyond.

TABLE 4.1. QUANLIA ($\mathcal{Q}$S) OF SWIMMEAN POWERS

No.	Power	ΔT		ΔQ (W)	Q (W)	Big-Bang Epoch
12	Collapse (last cycle)	$0 - \Lambda_S$	0–10^{-44} second	∞–10^{113}	∞	Planck
1	Seamlessness	$(1-\tau)\Lambda_S$	10^{-44}–10^{-39} second	10^{113}–10^{108}	10^{113}	Grand Unification
2	Centration	$(1-\tau)\tau\Lambda_S$	10^{-39}–10^{-34} second	10^{108}–10^{103}	10^{108}	Inflation
3	Allurement	$(1-\tau)\tau^2\Lambda_S$	10^{-34}–10^{-29} second	10^{103}–10^{98}	10^{103}	Electroweak
4	Emergence	$(1-\tau)\tau^3\Lambda_S$	10^{-29}–10^{-23} second	10^{98}–10^{92}	10^{98}	
5	Synergy	$(1-\tau)\tau^4\Lambda_S$	10^{-23}–10^{-18} second	10^{92}–10^{87}	10^{92}	
6	Transmutation	$(1-\tau)\tau^5\Lambda_S$	10^{-18}–10^{-13} second	10^{87}–10^{82}	10^{87}	
7	Transformation	$(1-\tau)\tau^6\Lambda_S$	10^{-13}–10^{-8} second	10^{82}–10^{77}	10^{82}	Quark
8	Cataclysm	$(1-\tau)\tau^7\Lambda_S$	10^{-8} second–1 ms	10^{77}–10^{72}	10^{77}	Hadron
9	Homeostasis	$(1-\tau)\tau^8\Lambda_S$	1 ms–1 fen*	10^{72}–10^{66}	10^{72}	Lepton
10	Interrelatedness	$(1-\tau)\tau^9\Lambda_S$	1 fen–1 year	10^{66}–10^{61}	10^{66}	Nucleosynthesis
11	Radiance	$(1-\tau)\tau^{10}\Lambda_S$	1 year–0.13 Ma†	10^{61}–10^{56}	10^{61}	Photon
12	Collapse	$(1-\tau)\tau^{11}\Lambda_S$	0.13 Ma–16.8 Ga‡	10^{56}–10^{51}	10^{56}	Entity

*1 fen = 4 minutes
†1 Ma = 1 million (10^6) years
‡1 Ga = 1 billion (10^9) years

In response to the twelve powers that drive the cosmo-ecological process at every stage, the evolution is characterized by a series of twelve spheres (Swimme 2004, 2017b; cf. Swimme and Anderson 2004). What follows is from the full unpublished manuscript on the powers of the universe, provided by Swimme (2017b). Most texts are quoted from the original, except for:

1. the texts reorganized under the numbered headings called "Sphere," a concept more accurately delineating cosmo-ecology;
2. the introduction of a new "power" (power 11), differentiated for complete stages of evolution; and
3. the correction of some typos and the updating of some data, for example, in describing the big-bang epochs, as listed in table 4.1.

Sphere 1

Driven by power 1, Seamlessness, this sphere starts at the end of the last cosmo-ecological cycle. The corresponding big-bang epoch falls in the slot of 0 to 10^{-39} s. This sphere evolves from the Planck Era (less than 10^{-44} s in time, less than 10^{-35} m in size, and greater than 10^{19} GeV in temperature at the end of the era*) at the end of the last cosmic cycle. The quantum-gravity wall has the stupendous initial energy to support the nonlocal connections that generate being, and the four fundamental gravitational, electromagnetic, strong, and weak interactions are merely one fundamental force at a unified quantum-gravity state. The epoch behaves as the primeval ground of being when the vacuum is not "pure" of nothingness but in a Grand Unification Era (less than 10^{-35} s, 10^{-27} m, greater than 10^{14} GeV). From $\approx 10^{-43}$ s gravity separates from the other three unified fundamental forces. It is the first instance of the breaking of symmetry among the forces in the quantum-gravity foam that appears and disappears too quickly to be measured, rising and falling in a vacuum within the Planck scale where things remain in an indeterminate state and all the possibilities (denoted by the physical wave function) are open.

In this sphere, there exist possibilities for all the things of the universe to be interconnected seamlessly with one another, beyond space and time, to form a widely dispersed but unitary reality, even though they seem to be far apart. Importantly, each thing in a certain sense begins as a bundle of possibilities before it actualizes into its form; in

*Data here and hereafter are cited from the tables in Ureview (2019) and Kazlev (2000).

other words, it emerges out of Seamlessness, and, as it emerges out of the gluey and interconnected realm of pure generativity, it increasingly refines itself into one particular definite being with an ongoing collapse of possibilities into one determination. Thus, Seamlessness acts as the initial stage of emptiness, which is permeated with the all-nourishing potential to generate being from nothing.

Sphere 2

Driven by power 2, Centration, this sphere corresponds to a self-centering epoch for real entities. The corresponding epoch is in 10^{-39} to 10^{-34} s. Within the epoch, the Inflation Era initiates from $\approx 10^{-36}$ s (less than 10^{-32} s, 10^{-24} m, greater than 10^{14} GeV). At $\approx 10^{-35}$ s, strong force freezes out from the unified electroweak force through the second-phase transition, and big bang occurs with a cataclysm to drive cosmic inflation in space in order of 10^{26} within a time of $10^{-35 \approx -34}$ s, and the universe is spatially isotropic and homogeneous.

The power of the sphere concerns the creativity when the universe centers on itself in its early stage of primeval particles. When the universe is billowing out to increase the gravitational potential energy transferred from the energy or heat, the centering initiates and results in omni-centricity because the process happens in each being everywhere in the universe. Looking at the universe from any particular position, one finds that the universe is expanding from each of those positions. That is, every position is the center of the expansion. The fact that no matter what place we look at, the cosmic expansion proceeds from that place demonstrates the way that the universe is centering upon itself everywhere. It is the isotropic and homogeneous property of the universe, the so-called cosmological principle.

In the sphere, every being is the center in its particular way, just as in the earth community, every species has its particular significance as a center in the ecological web. However, the centering process of creativity does not remain stable while the universe is expanding. A multiplicity of centers form when the expanding universe fractures apart and individual galaxies emerge from primeval particles. Every galaxy becomes a new creative center of the universe. The centering process occurs in such a way that every center unites with other centers in the universe. Every unification gives rise to deeper creativity, the depth of which depends upon that of the Centration, while the creativity is only possible among beings that are involved with Centration. In Earth's ecological system, similar processes of creativity move toward centers that eventually reach the sophisticated state that we call a cell.

Sphere 3

Driven by power 3, Allurement (or attraction), this sphere holds entities together. The corresponding epoch lasts from 10^{-34} s to 10^{-29} s. The Electroweak Era starts at 10^{-32} s and continues to 10^{-12} s ($< 10^{-11}$ s, 3 mm, $>$ 300 GeV). The Inflation Era ends at $\approx 10^{-32}$ s with energy transformed into matter, but the expansion continues at a lesser rate. Bosons are formed rapidly, far from the thermal equilibrium and driving a preheating process. A slight excess (10^{-9}) of matter over antimatter develops to give the presently observed predominance of matter over antimatter. Dark matter (such as axions) may begin to form.

In a broader view, the power of the sphere is evident in sexuality, another typical way in which Allurement enters into an intimate relationship among entities. Sexual beings are two organisms that can join with each other to create offspring. This creativity began around 2 billion years ago, and over this period of sexual history, there has been an ongoing and deepening journey into greater intimacy. In terms of sexuality over the last 500 million years, not only do these organisms fit together but when getting to advanced organisms with backbones, they find a way to deepen their sexual relationship with the development of sexual organs and the genitalia. The reptiles found a way to evolve the penis and the vagina to deepen into intimacy the journey that begins with Allurement. These developments led to deep sexual arousal that can be coded, maintained, and remembered genetically. By contrast, Allurement in terms of humans is best expressed by sexual attraction. It is domesticated and controlled to behave as both the cause and effect of ecstasy that permeates the intimacy of mutual enhancement between two beings through the effect of the pleasure and joy that one feels while knowing that his or her own pleasure and joy is a cause of the other's pleasure and joy.

Sphere 4

Driven by power 4, Emergence, this sphere accommodates the continually arising of self-assembly and self-organization. The corresponding epoch is 10^{-29} to 10^{-23} s, accompanied by the continued Electroweak Era. Exotic particles like W/Z and the 125-GeV Higgs bosons appear due to the spontaneous symmetry breaking. The Higgs field has two parts: a dynamic field with quanta Higgs bosons, and a constant one (called the vacuum expectation value), proportional to the mass of other fields (interpreted as giving mass in quantum field theory). Elementary particles (except photons, gluons, and three generations of neutrinos) begin to gain mass in the Higgs field, in contrast to composite particles (protons, nuclei, atoms) the mass of which come from the binding energy that holds these particles together. Elementary particles can mix and combine into more elementary particles.

In cosmology, the universe is found to go through the great birth moments of creativity, collectively known as Emergence. The moments represent both self-assembly and self-organization. There are several creative moments when the power of Emergence shows up. The first epochal moment happens when the universe experiences a flaring forth, the primeval fireball. The second arises when the first stars or galaxies pull themselves out of the whole vast uniform expansion of the universe by overcoming the pressure of Allurement to continue expanding. The third comes when stars and galaxies release themselves into their creativity by the production of carbon, the basis of all life, at the center of stars. The fourth appears with the birth of planets, at least one of which becomes alive to endure a series of self-assembling breakthroughs for advanced life to show up. The fifth exists in the whole history of life when nuclei-equipped eukaryotic cells evolve from a symbiotic association of the nucleus-free prokaryotes. The last took place when humans, collectively, gained the ability to influence and shape the Earth's geological processes. Naturally, when the human species emerged as a significant presence, on par with the geological evolution of Earth's biosphere, the Earth underwent a transition from the preceding geological Cenozoic Era to a new Ecozoic era. This transition may extend to the ecological enhancement of additional planets such as Mars.

Sphere 5

Driven by power 5, Synergy, this sphere gives the epoch of 10^{-23} to 10^{-18} s. The Electroweak Era continues in which the temperature (> 100GeV) is too high for quarks to remain clumped to coalesce into hadrons, instead to form a quark-gluon plasma (QGP) or quark soup. All the known particles (including quarks, leptons, gluons, photons, Higgs bosons, and W and Z bosons) were extremely relativistic, with the highest energies able to be directly observed in an experiment in the Large Hadron Collider.

Synergy means working together, or creating together, or doing together, or being together, or living together. All of these happen related to the universe. There could be various kinds of Synergy, but the positive one is when a relationship leads to a functional effect that is beneficial and would not be there otherwise than in the relationship itself. A good example is the deep cooperative interaction among large groups of penguins to survive on the South Pole during the intense winters. Another example is the synergistic relationship between plants that need nitrogen and bacteria or fungi that need sugar. Trees provide sugar to enable bacteria to live, while the bacteria form nodules to fix the nitrogen out of the atmosphere into the plant. The same happens for the synergistic process of pollination between plants and insects, photosynthesis between plants and sunlight, cooperation between some squid and bioluminescent bacteria, the non-decaying structure between neutrons and protons, and sexuality between male and female organisms to face and overcome the challenges of food and offspring.

The conscious self-awareness of Synergy is mandatory to meet the challenges of our time. This period is a new transition moving from industrial civilization to a planetary one, from one form of humanity to a new form, and, enhanced by the synergistic qualities of mind and consciousness, a broadened ecological system accompanied by a growth in the size and complexity of biological structures.

Sphere 6

Driven by power 6, Transmutation, this sphere corresponds to the epoch of 10^{-18} to 10^{-13} s, the last stage of the Electroweak Era until 10^{-12} s. The electromagnetic and weak interactions separate from each other, and the electroweak boson breaks into the photon of electromagnetism and three bosons of the weak force (two charged W bosons and a neutral Z boson), and the W and Z bosons eventually cease to form. Unstable massive particles disappear when the temperature falls to a value at which photons from the black-body radiation do not have sufficient energy to create a particle-antiparticle pair. Baryogenesis and the decoupling of cold dark-matter particles come into existence.

The same as other powers, Transmutation pervades the universe and permeates every cell, every organ, and every organism. It exists where elementary formless plasma particles go through changes into the form of atoms. It exists in the vast clouds of atoms where galaxies give birth and within which the primal stars develop into stellar systems with planets. It exists on the Earth where molten rocks form a living planet on the surface of which simple primeval cells evolve into prokaryotes, into eukaryotes, into eventually the advanced forms of ecosystems, part of which is the human.

Human beings participate in this transmutation from the Cenozoic Era to the Integral Era, also called the Planetary Era or the Ecozoic Era, through the creation of constraint, resistance, judgment, and even rejection, embodied in rules, laws, customs, disciplines, religions, and cultures. The process leads to, on the one hand, the protection of individual rights; on the other hand, the safeguarding of the whole community of life, including humanity, allowing it to prosper and thrive. The shift leads to the excellence of the dynamic natural selection throughout every aspect of the ecosystems by deepening the intimacy among individual species and guaranteeing the integrity of the whole evolving community.

Sphere 7

Driven by power 7, Transformation, the sphere defines the epoch in 10^{-13} to 10^{-8} s, featured by the Quark Era of 10^{-12} to 10^{-6} s (at 10^{-6} s: 1.2×10^{13} K, 1.4 light-days). At 10^{-11} s, weak and electromagnetic forces split through the last phase transition and the now-distinct four fundamental forces. The gravity starts to dominate expansion. Leptons separate into electrons, neutrinos, and antiparticles, and quarks become confined together to create matter primarily formed from the individual building blocks of protons, neutrons, electrons, neutrinos, and their antiparticles.

Differing from the previous sphere, this sphere is necessary for the entire context encompassing life or society as a whole. It provides an individual-level adaptation within the sphere and imposes a requirement for personal change. The resultant force inherent in this sphere propels the individual's transformative process, integrating it into a broader context.

From the perspective of the universe, the evolution of the whole in different epochs is upgraded by the evolving individuals at different stages. The very early moment of the universe is represented by the early plasma state in which elementary particles like electrons then protons become stable structures and endure through time. The structuring of the universe takes place through a gathering of the different moments at which stars and galaxies are formed to open the possibility for further development. Even more significantly, the harmonious relationship that has taken place throughout time becomes recognizable only after the appearance of the human cultural ability to store learning outside the body, and thus the universe becomes more intense in its folding back upon its earlier breakthrough moments (i.e., its space-time binding).

That is to say, because of the human macrophase power, its cultural coding becomes the recoding of the planet upon which humanity dwells; and no doubt, it is the recording of the universe from which the planet originates. Therefore, it is particles, stars, galaxies, and at present, humans—that is, all the individual entities of the universe—that enable the universe to go forward in its work of space-time binding of the past into the present, with more of the universe present to consciousness now than at any time previously.

Sphere 8

Driven by Power-8, Cataclysm, this sphere destructs entities, albeit the ones too significant to sustain the existing system. The corresponding epoch is in a duration of 10^{-8} s to 1 ms, within which the Hadron Era starts from 10^{-6} s and lasts till 1 s (10^{10} K, four light-years). In the epoch, quarks are bound into hadrons like protons and neutrons. A slight matter-antimatter asymmetry from the earlier phases (baryon asymmetry) results in an elimination of anti-hadrons. Electrons collide with protons to fuse to form neutrons and give off massless neutrinos, which continue to travel freely through space today, at or near to the speed of light. Some neutrons and neutrinos recombine into new proton-electron pairs. Conservations in total charge and energy (including mass-energy) are the only rules to govern all the random combinations and recombinations. The universe takes shape $\approx 10^{-6}$ s with a slow expansion, cooling down, and density decreases. The most fundamental forces take effect to hold the nuclei of atoms together: at first gravity, then the strong force, followed by the weak and electromagnetic forces.

Cataclysm is related to the idea of the things of macrocosmic scales breaking down from the complex into the simple. In physics, it is referred to as the most fundamental law, the second law of thermodynamics that breaks entities down so new beings can come forth. An example to show the existence of Cataclysm is given by stars that are twenty times the size of the Sun. For ten million years, they burn out hydrogen constantly to build up helium. They then burn and compact helium into carbon in the form of the massive diamond called a white dwarf. Such things also exist when more giant stars crush the diamond down further until the carbon begins to burn for oxygen, for silicon, and for iron at the end, leading to the formation of pulsars or neutron stars. Cataclysm is what the universe is doing to break down complex objects moment after moment.

On our planet Earth, there happened another example of Cataclysm. It was the forty-seven-million-year mass extinction at the end of the Permian (298.9–251.9 Ma years ago) in the last period of the Paleozoic era. Ninety-six percent of marine species plus perhaps 70 percent of the land species were obliterated. The extinction almost escaped the attention of historical records until the nineteenth century; yet, it did not escape the power of the universe whether or not we intellectually know about the cataclysm and the empirical evidence for the mass extinction.

Sphere 9

Driven by power 9, Homeostasis, this sphere is the Lepton Era (1 s to 3 minutes; 200 s: 8.4×10^8 K, 55 light-years). As the temperature drops, quark-antiquark annihilation stops, and the remaining quarks combine to make protons and neutrons. By 1 s, the universe is made up of fundamental particles and energy: quarks, electrons, photons, neutrinos, and less familiar types. These particles smash together to form protons and neutrons. At ≈1 s, neutrinos and antineutrinos decouple from interactions and become inactive. The electrons and positrons are annihilated and are not recreated. An excess of electrons is left. The neutron-proton ratio shifts from 50:50 to 25:75. At ≈15 s, electrons and positrons annihilate each other to give more photons. A slight excess of electrons is left but is the same as the number of protons to keep zero net charges. Leptons and anti-leptons remain in thermal equilibrium. During ≈3 s to 2 minutes, initial nucleosynthesis happens allowing fundamental elements to form: protons and neutrons come together to form primordial nuclei of hydrogen, helium, and lithium; but it will take another 300,000 years for electrons to be captured into orbits around the nuclei to form stable atoms.

As the energy to maintain the development of the universe,

Homeostasis distributes over the whole system. It illustrates the dynamics that maintain the form and function of a mammal's body. It corresponds to the human bloodstream where the pH balance is the same as in the bloodstreams of most animals and fish. It is revealed in the temperature of the Earth, which remains in a state where life can flourish, though in the presence of the cooling-warming cycles of about 100,000 years. It characterizes the Milky Way supernovae, the explosion of which fluctuate alternatively between 8,000 and 12,000 times every 1 million years. Also, it manifests in the expansion of the universe as a whole, moving apart in its initial explosion and yet holding itself together with its gravitational attraction. In the process, the expansion perfectly balances the attraction. Thus, there is no way for the universe either to expand apart into a thinner one or to collapse into a denser one. It is just at an expanding rate, which is necessary to push forward the blossom of particles, stars, and galaxies for the birth of beings and lives within the whole single body of the universe.

Sphere 10

Driven by power 10, Interrelatedness, this sphere spans the epoch from 4 minutes to 1 year. It is the Nucleosynthesis Era (3–20 minutes; 1,000 s: 4×10^8 K, greater than 55 light-years) followed by the pre-phase of the Photon Era (3 minutes to 1 year). The initial nucleosynthesis of ≈3 s to 2 minutes goes on to ≈3 minutes at which heavier nuclei, such as deuterium, are formed under conditions similar to those in stars today or in thermonuclear bombs. With a time constant of ≈1,000 s, neutrons are used up, and any remaining ones decay. The neutron-proton ratio comes to 13:87. The bulk constitution of the universe is now in place consisting mainly of 75 percent protons and 25 percent helium nuclei. The temperature is still too high to form any atoms because electrons behave as a gas of free particles. After ≈20 minutes, the temperature and density of the universe have fallen to the point where nuclear fusion cannot continue. From ≈3 minutes, the Photon Era begins because temperatures remain too high for the binding of electrons to nuclei, and the universe starts to exist in the plasma state, consisting of atomic nuclei, electrons, and photons.

In contrast to the power of Allurement, the attraction toward the self, Interrelatedness is the care that the universe develops for 13.8 billion years to keenly attune to relationships. It is a capacity to identify the worth and meaning of being in a relationship with another. Note that there exists a distorted or negative form of this power: the loss of self when absorbed in the needs and the values of others. In this case, a submissive (or dependent) relationship forms, and thus Centration is necessary to build up the self.

One of the most prominent examples of Interrelatedness is parental care. The love of a parent makes the survival of offspring more likely. It is just this kind of care that has survived throughout the biosphere for hundreds of millions of years on microcosmic Earth. In the biosphere, humanity appears to be "the product of a seemingly endless evolution within a loving plan of plurality, union and a resultant energy transformation into something more complex, more conscious and more loving from the pre-atomic stage in the depths of the primordial sea to his present hominized state" (Roland 2009). Similarly, our universe has been ubiquitously expanding overwhelmingly for 13.8 billion years at the rate that leads to life. It is the macrocosmic parental care that earlier humans symbolized as the Great Mother that is built in from the beginning of the universe.

Sphere 11

Driven by power 11, Radiance, this sphere allows entities to express their magnificence into the universe at both a surface and a depth level of perception. It is the celebratory revelation of the profound mystery of the cosmos. Everything is a source of this power. For humans, it is the capacity to respond to the radiative energy filling the universe as well as the capacity to respond to that within. It is most commonly expressed as the final smile on the face of a dying human or the supernova explosion of a dying star.

In cosmology, the corresponding epoch starts from 1 year to 0.13 Ma. It is the radiation-dominated Photon Era (1–240,000 years; in the end, 7,000 K, less than 2×10^5 light-years) after most of the leptons and antileptons have annihilated each other at the end of the Lepton Era. The universe is at the plasma state with most of its energy coming from the remnant of the primordial fireball to be dominated by photons in the form of radiation in different wavelengths of light, X-rays, radio waves, and ultraviolet rays. It is the first significant era in the history of the universe.

From the beginning of time, radiant energy has coursed through the universe. All the powers are attempting to bring forth radiance, the primary and primal way of the universe to express beauty. It has two steps. The first step is the surface-level resonance with the radiance in the universe. However, if the resonance is deep enough, what comes forth is a reverberation at an in-depth level. When the reverberation begins, a non-dual relationship is set up between the entity giving expression to beauty and the entity receiving this expression. In traditional terms, resonance is the primary form of prayer, and reverberation is the primary sacrament in the universe. While the former enters into contemplation to find a way to enter deeply into things, the latter is the way to become the very radiance flooding the world.

Sphere 12

Driven by power 12, Collapse, this is the sphere hidden behind the Spheres driven by power 1, Seamlessness, power 6, Transmutation, and power 8, Cataclysm. It is the annihilation sphere where nothing continues to exist, yet a new universe may be reemerging from quantum fluctuations to give rise to the Seamlessness stage for the next cycle. It is the final epoch that lasts from 0.13 Ma to 16.8 Ga. The ESA's Planck Mission reported that the age of the present universe is 13.799 ± 0.021 Ga (Planck Collaboration 2016, last column in table 4: Age/Gyr).

According to a recent work (Jiménez et al. 2016), our universe should remain from now on an additional 2.8 Ga, the lowest limit of its cosmic evolution. Thus, its total age may be more than 16.6 Ga. This limit is consistent with a full big-bang cycle of 16.8 Ga with a margin of error of 1.2 percent. It is worth noting that, although assumptions with much longer ages (e.g., from 10^{14} to 10^{100} years) were discussed in either a cosmic "big rip" (Caldwell et al. 2003) or a cosmic "big freeze" (Bohmadi-Lopez et al. 2008), such models lacked either theoretical credibility or experimental supports, and thus should be considered nonviable at the present time (Adams and Laughlin 1997).

SUMMARY

Swimme's evolutionism scientifically transcends the theological cosmic evolution framework proposed by Teilhard and Berry. He adeptly integrates and reconciles spiritual beliefs with a wide array of scientific perspectives, extending beyond evolutionary biology to encompass broader scientific narratives such as big-bang cosmology. His synthesis forms a comprehensive scientific paradigm, anchored in the concept of twelve cosmogenetic powers that act as fundamental drivers orchestrating the intricate dance of all elements within the cosmos. This dance unfolds in a complete cycle, reflecting a purposeful and guided evolution that goes beyond the mere biological realm. Swimme's vision suggests a cosmo-ecological narrative intricately woven with deeper spiritual implications, portraying the universe as a dynamic and interconnected whole propelled by these powers.

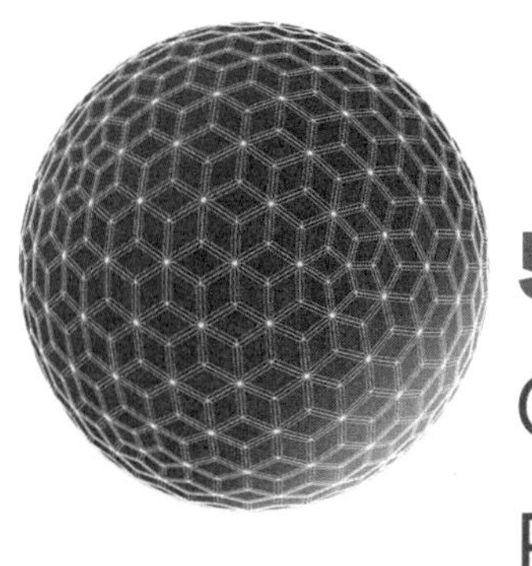

5 Quantitative Principles of the Holographic Universe

THE COALESCENCE OF EASTERN WISDOM AND SWIMMEAN EVOLUTIONISM

In the previous chapter, we discussed that, rooted deeply in the Eastern philosophy, Teilhard proposed the concept of God-Omega, a Christian term that nonetheless expresses both the ontological and the cosmological essence of *dao*. In addition, Berry borrowed taiji's *yin-yang* properties to annotate the mutual coupling between the physical part and the psychic part of the "radial energy." Furthermore, going beyond Teilhard-Berry's Christian framework, Swimme synthesized scientific achievements to identify and systematize a set of twelve cosmogenetic powers. These powers reveal the sacred wholeness of reality, intricately weaving processes from the beginning of time to the appearance of matter, elements, galaxies, and life. They are organically coupled within twelve epochal spheres, respectively, active across all levels and scales (Amberg 2011, pp. 7–39; Baker 2015, pp. 1–7; Le Grice 2011, pp. 226–38).

Examining the four features of Swimmean epochal powers—closed

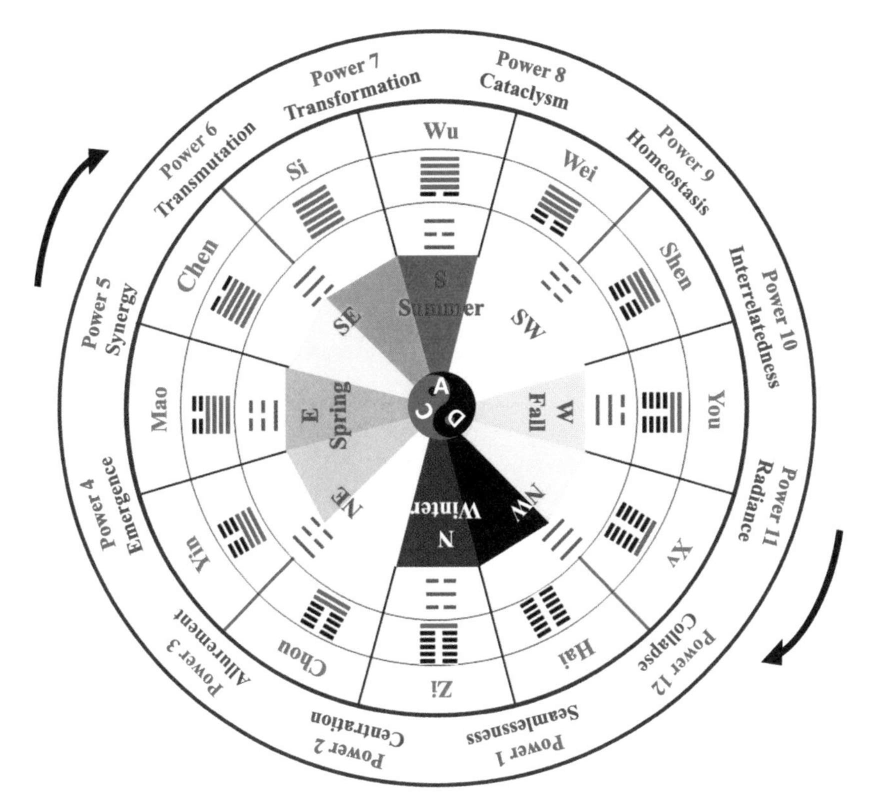

Figure 5.1. Correspondence of Swimmean powers and the I Ching's sovereign hexagrams

interdependence, pair reflexivity, pair complementarity, and cause-effect relations—we see that the twelve elements have a one-to-one correspondence to the I Ching's twelve sovereign hexagrams of the *qi* elements. They demonstrate a complete cycle of the materialization of the universe that grows out of Seamlessness and disappears into the next Seamlessness. Figure 5.1 illustrates the correspondence in a complete cycle.

For convenience of comparison, let us first rewrite here the dependent-arising chain of the twelve Swimmean powers: P1-Seamlessness → P2-Centration → P3-Allurement → P4-Emergence → P5-Synergy → P6-Transmutation → P7-Transformation → P8-Cataclysm → P9-Homeostasis → P10-Interrelatedness → P11-Radiance → P12-Collapse → P1-Seamlessness (next cycle).

This chain closely parallels the dynamic process described by ancient Daoist onto-cosmology, exhibiting striking similarities to the trend of everything emerging from the prior Nothingness (0 ䷁) and then dissolving into the next Nothingness through a complete cycle. The cycle can be delineated into a series of stages not only by the twelve sovereign hexagrams but also by the twelve terrestrial branches: H12-Receptivity 0 ䷁ hai → H1-Returning 32 ䷗ zi → H2-Overseeing 48 ䷒ chou → H3-Tranquility 56 ䷊ yin → H4-Great Strength 60 ䷡ mao → H5-Resoluteness 62 ䷪ chen → H6-Creativity 63 ䷀ si → H7-Meeting 31 ䷫ wu → H8-Retreat 15 ䷠ wei → H9-Divorcement 7 ䷋ shen → H10-Correspondence 3 ䷓ you → H11-Disintegration 1 ䷖ xv → H12-Receptivity 0 ䷁ hai (next cycle). Table 5.1 provides a detailed overview of correspondence, integrating with figure 4.5 (p. 94).

P1-Seamlessness is the first epochal power responsible for the transition from the terrestrial branch B12 hai of the prior cycle to B1 zi of the present cycle, symbolized by the I Ching's two sovereign hexagrams, H12 Receptivity 0 ䷁ and H1 Returning 32 ䷗, respectively. It initiates B1 zi and ends at the middle of Zi. It fills the nebulous voids and drives a cyclic return from the prior cycle.

P2-Centration is the second epochal power responsible for the transition from B1 zi to B2 chou, symbolized by H1 Returning 32 ䷗ and H2 Overseeing 48 ䷒, respectively. It initiates B2 chou and ends at the middle of chou. It presides over the initial movement of the primordial *qi* from nonbeing to being in the voids, a stage of the Great Easiness.

P3-Allurement is the third epochal power responsible for the transition from B2 chou to B3 yin, symbolized by H2 Overseeing 48 ䷒ and H3 Tranquility 56 ䷊, respectively. It initiates B3 yin and ends at the middle of yin. It gives the primordial *qi* to emerge ubiquitously but without form, a stage of the Great Origin.

TABLE 5.1. LIST OF SWIMMEAN POWERS AND THE I CHING'S SOVEREIGN HEXAGRAMS

Sovereign Hexagram				Swimmean Power				
H12-Receptivity (prior)	0	䷁	B12-hai	P1-Seamlessness	Past	Communion (C)	Cause 1	Output
H1-Returning	32	䷗	B1-zi					Input
H2-Overseeing	48	䷒	B2-chou	P2-Centration		Autopoiesis (A)	Effect 1	Action
H3-Tranquility	56	䷊	B3-yin	P3-Allurement	Present			
H4-Great Strength	60	䷡	B4-mao	P4-Emergence				
H5-Resoluteness	62	䷪	B5-chen	P5-Synergy				
H6-Creativity	63	䷀	B6-si	P6-Transmutation		Differentiation (D)	Cause 2	Output
H7-Meeting	31	䷫	B7-wu	P7-Transformation				Input
H8-Retreat	15	䷠	B8-wei	P8-Cataclysm			Effect 2	Action
H9-Divorcement	7	䷋	B9-shen	P9-Homeostasis	Future			
H10-Correspondence	3	䷓	B10-you	P10-Interrelatedness		Communion (C)		
H11-Disintegration	1	䷖	B11-xv	P11-Radiance				
H12-Receptivity	0	䷁	B12-hai	P12-Collapse				
H1-Returning (next)	32	䷗	B1-zi	P1-Seamlessness (next)	Past (next)		Cause 1	Output

P4-Emergence is the fourth epochal power responsible for the transition from B3 yin to B4 mao, symbolized by H3 Tranquility 56 ䷊ and H4 Great Strength 60 ䷡, respectively. It initiates B4 mao and ends at the middle of mao. It energizes the emergence of form from the primordial *qi* in the absence of matter, known as the stage of the Great Inaugural.

P5-Synergy is the fifth epochal power responsible for the transition from B4 mao to B5 chen, symbolized by H4 Great Strength 60 ䷡, and H5 Resoluteness 62 ䷪, respectively. It initiates B5 chen and ends at the middle of chen. It forms the stage of the

Great Simplicity, where entity-free matter separates and is distinct from the primordial *qi* and and form producing the Chaos of the original *qi*.

P6-Transmutation is the sixth epochal power responsible for the transition from B5 chen to B6 si, symbolized by H5 Resoluteness 62 ䷪ and H6 Creativity 63 ䷀, respectively. It initiates B6 si and ends at the middle of si. It is for the birth of the stage of the Great Ultimate (taiji) by constantly fueling the original *qi* beyond the Ultimateless (wuji).

P7-Transformation is the seventh epochal power responsible for the transition from B6 si to B7 wu, symbolized by H6 Creativity 63 ䷀ and H7 Meeting 31 ䷫, respectively. It initiates B7 wu and ends at the middle of wu. It drives the two polar lines, *yin* and *yang*, to nestle together and be mutually responsive to each other in the Oneness of the Great Ultimate (taiji).

P8-Cataclysm is the eighth epochal power responsible for the transition from B7 wu to B8 wei, symbolized by H7 Meeting 31 ䷫ and H8 Retreat 15 ䷠, respectively. It initiates B8 wei and ends at the middle of wei. It energizes four digrams and five phases to an overwhelming stage of the *yin-yang* energy redistribution in larger scales.

P9-Homeostasis is the ninth epochal power responsible for the transition from B8 wei to B9 shen, symbolized by H8 Retreat 15 ䷠ and H9 Adversity 7 ䷋, respectively. It initiates B9 shen and ends at the middle of shen. It gives rise to eight trigrams in a stabilized and unceasing process to further separation of *qi* in the four digrams. This is a crucial stage of differentiation (D).

P10-Interrelatedness is the tenth epochal power responsible for the transition from B9 shen to B10 you, symbolized by H9 Adversity 7 ䷋ and H10 Correspondence 3 ䷓, respectively. It initiates B10 you and ends at the middle of you. It strengthens the interrelationships of entities through adjusting mutual correspondences for a more prolific stage of sixty-four hexagrams.

P11-Radiance is the eleventh epochal power responsible for the transition from B10 you to B11 xv, symbolized by H10 Correspondence 3 ䷓ and H11 Disintegration 1 ䷖, respectively. It initiates B11 xv and ends at the middle of xv. It is for the 64 × 64 square-circle stage that communicates in deepened interrelationships, however, via a self-decaying or self-exhausting way so as to reach physical partings.

P12-Collapse is the epochal power responsible for the transition from B11 xv to B12 hai, symbolized by H11 Disintegration 1 ䷖ and H12 Receptivity 0 ䷁, respectively. It initiates B12 hai and ends at the middle of hai. It is for the root-returning stage of the ongoing cycle toward nothingness by obediently receiving the power's domination.

In a nutshell, table 5.2 (p. 116) presents the coalescence of this account of Swimmean evolutionism and Eastern wisdom.

The table starts from a direct comparison between Teilhard's panentheistic trinity-power philosophy and Daoist atheistic *li-qi* ideology. It then expresses the ontological wholeness of reality, stressing the cosmological congruence between nature and humanity. Finally, it constitutes the correlations between Swimmean powers and corresponding hexagrammatic *qi* that dominates twenty-four solar terms.

The congruence between the elements of the evolutionism and the hexagrams outlined in table 5.2 implies that the powers are in phase with hexagrammatic *qi* and staggered with all the solar terms throughout the cycle. For example, P1 Seamlessness covers the medial term, Less Snow, and the next nodal one, Great Snow; P2 Centration drives the medial term, Winter Solstice, and the nodal one, Less Cold; and so on.

Considering that *qi* is a kind of (vital) power, it is natural to accept that the Swimmean powers are intrinsically identical to the hexagrammatic *qi*, albeit in alternative sets of nomenclatures. Yet it deserves mention that the two sets are 90 degrees out of phase with each other—that is, when one is at its peak, the other is at its trough.

TABLE 5.2. COALESCENCE OF SWIMMEAN EVOLUTIONISM AND EASTERN WISDOM

Paradigm	Swimmean Evolutionism	Eastern Wisdom
Reality	Sacred universe	Heaven–Humanity Oneness
Ontological Body	God-Omega	*Dao* / Wuji
Ontological Principle	Cosmogenetic trinity	*Li* / Taiji
Cosmological basis	Physical & psychic dimension	*Qi* Distribution
Prime mover	Swimmean powers	Hexagrammatic *Qi*
Onto-cosmology	Trinity-Power	*Li* - *Qi*

Power	*Qi*				Solar term	
P1-Seamlessness	Nebulous void		䷗ Zi	Returning	Nodal	Great Snow
P2-Centration					Medial	Winter Solstice
	Great easiness	Wuji	䷒ Chou	Overseeing	Nodal	Less Cold
P3-Allurement					Medial	Great Cold
	Great origin		䷊ Yin	Tranquility	Nodal	Spring Beginning
P4-Emergence					Medial	Rain Water
	Great inaugural		䷡ Mao	Great Strength	Nodal	Insect Waking
P5-Synergy					Medial	Spring Equinox
	Great simplicity		䷪ Chen	Resoluteness	Nodal	Pure Brightness
P6-Transmutation					Medial	Grain Rain
	Great ultimate	Taiji	䷀ Si	Creativity	Nodal	Summer Beginning
P7-Transformation					Medial	Less Fullness
	Two monograms		䷫ Wu	Meeting	Nodal	Grain in Ear
P8-Cataclysm					Medial	Summer Solstice
	4-xing & five phases		䷠ Wei	Retreat	Nodal	Less Heat
P9-Homeostasis					Medial	Great Heat
	8-gua		䷋ Shen	Adversity	Nodal	Autumn Beginning
P10-Interrelatedness					Medial	Heat End
	64-gua		䷓ You	Correspondence	Nodal	White Dew
P11-Radiance					Medial	Autumn Equinox
	64×64 square-circle		䷖ Xu	Disintegration	Nodal	Cold Dew
P12-Collapse					Medial	Frost Descending
	Root-returning		䷁ Hai	Receptivity	Nodal	Winter Beginning
P1-Seamlessness					Medial	Less Snow

Besides, there exists a correlation between the six-pair reflective powers and six-pair waning-waxing hexagrammatic *qi* elements. That is to say, the two powers of every pair have the same reflective property with each other, exactly the same as the two-adjacent hexagram *qi* elements:

P1-Seamlessness ䷁ → ䷗ ↔ ䷖ → ䷁ P12-Collapse
P2-Centration ䷗ → ䷒ ↔ ䷓ → ䷖ P11-Radiance
P3-Allurement ䷒ → ䷊ ↔ ䷋ → ䷓ P10-Interrelatedness
P4-Emergence ䷊ → ䷡ ↔ ䷠ → ䷋ P9-Homeostasis
P5-Synergy ䷡ → ䷪ ↔ ䷫ → ䷠ P8-Cataclysm
P6-Transmutation ䷪ → ䷀ ↔ ䷀ → ䷫ P7-Transformation

Furthermore, there are six complementary (mirror) power-*qi* pairs. This means that the two powers among the twelve ones are also complementary, the same as that of the two hexagrammatic *qi* elements:

P1-Seamlessness ䷁ → ䷗ ↔ ䷀ → ䷫ P7-Transformation
P2-Centration ䷗ → ䷒ ↔ ䷫ → ䷠ P8-Cataclysm
P3-Allurement ䷒ → ䷊ ↔ ䷠ → ䷋ P9-Homeostasis
P4-Emergence ䷊ → ䷡ ↔ ䷋ → ䷓ P10-Interrelatedness
P5-Synergy ䷡ → ䷪ ↔ ䷓ → ䷖ P11-Radiance
P6-Transmutation ䷪ → ䷀ ↔ ䷖ → ䷁ P12-Collapse

It is worth mentioning that, in the archetypal cosmos (Le Grice 2011, pp. 222–23), the twelve powers act exactly on the twelve divisions of the solar ecliptic longitude. These divisions are expressed by the twelve ecliptic signs: Aries (♈), Taurus (♉), Gemini (♊), Cancer (♋), Leo (♌), Virgo (♍), Libra (♎), Scorpio (♏), Sagittarius (♐), Capricorn (♑), Aquarius (♒), and Pisces (♓).

QUANTIFICATION

Eastern culture nurtured world-leading achievements in ancient China, like gunpowder, the compass, printing, silk making, and the first book (that is, the I Ching). During the Xia dynasty, the lunisolar calendrical system became available. It rested upon the five pillars related to human lives: Year (nian), Month (yue), Day (ri), Double Hour (shichen), and Four Minutes (fen), where 1 Year (nian) contains 129,600 elements of fen. Simplifying 129,600 elements of fen as 1 great fen (dafen), we have 1 Year (nian) = 1 great fen (dafen).

During the Song dynasty (960–1279 CE), Shao Yong discovered a structured world-ordering principle by generalizing the lunisolar calendrical system, guided by the I Ching's onto-cosmology. The principle was applied to a 129,600-year cycle that is in the order of a terrestrial ice-age period on the Earth. The constructed numero-cosmological catalog logically amplified the five human pillars by a factor of 129,600 to the following five terrestrial ones, namely: Cycle (yuan), Epoch (hui), Revolution (yun), Generation (shi), and Year (nian).

The evolving stages of both systems were found to fit with the same dimension-free hexagrammatic set of the twelve sovereign hexagrams of the qi elements. This numeral synopsis implies that, at least within the ice-age scale of 129,600 years, nature is holographic and holofractal. Consequently, we are able to rely on the known properties of individual entities to investigate, understand, interpret, and/or predict the cosmo-ecological evolution of a world beyond the ice-age scale.

Such a world is the observable universe. It is at the cosmological age of about 13.8 billion years. Fortunately, the sacred wholeness of this larger-scale reality has well been elucidated by the Swimmean evolutionism transcending the Teilhard-Berry frame. The suggested twelve epochal powers match up with the I Ching's twelve sovereign hexagrams. Therefore, it is promising to explore the quantitative principles of natural things that give rise to a holographic unity among the cosmic heaven, the planetary Earth, and conscious humankind.

Three Tiers of Holography

We have identified a couple of holographic tiers. The first one is the initial microcosmic tier, built upon five pillars associated with human lives. The second one is the mesocosmic tier, connected to terrestrial ice ages. Undoubtedly, we can employ a similar methodology to establish a new tier. This third tier aims at providing macrocosmic pillars for probing into the celestial level of the universe. Taking into account the factor of 129,600 that exists between the five pillars of the two known tiers, we derive the five celestial pillars by amplifying the terrestrial ones with the same factor, resulting in the following:

- Great Cycle (*dayuan*)
- Great Epoch (*dahui*)
- Great Revolution (*dayun*)
- Great Generation (*dashi*)
- Great Year (*danian*), which is nothing else but the Cycle (yuan), i.e.,

1 Cycle (yuan) = 1 Great Year (danian)

Thus, we have:

- **Tier 1, five human pillars:** Year (nian), Month (yue), Day (ri), Double Hour (shichen), and Four Minutes (fen)
- **Tier 2, five terrestrial pillars:** Cycle (yuan), Epoch (hui), Revolution (yun), Generation (shi), and Year (nian) = great fen (dafen)
- **Tier 3, five celestial pillars:** Great Cycle (dayuan), Great Epoch (dahui), Great Revolution (dayun), Great Generation (dashi), and Cycle (yuan) = Great Year (danian)

For a concise description, Chinese pinyin is used hereafter to express the nomenclature. A close look at the above five pillars of the three tiers leads to the conclusion that the three tiers are correlated, establishing the following relations:

First, there is a factor of 129,600 existing among the three tiers:

- 1 dayuan = 129,600 yuan, while 1 yuan = 129,600 years
- 1 dahui = 129,600 hui, while 1 hui = 129,600 months
- 1 dayun = 129,600 yun, while 1 yun = 129,600 ri
- 1 dashi = 129,600 shi, while 1 shi = 129,600 shichen
- 1 yuan = 1 danian = 129,600 dafen
- 1 year = 1 dafen = 129,600 fen

Second, every tier has the same category of twelve and thirty coefficients:

- 1 year/yuan/dayuan = 12 months/hui/dahui
- 1 month/hui/dahui = 30 ri/yun/dayun
- 1 ri/yun/dayun = 12 shichen/shi/dashi
- 1 shichen/shi/dashi = 30 fen/years/yuan

Finally, there exist the relations of

- 1 year/yuan/dayuan = 12 months/hui/dahui = 360 ri/yun/dayun = 4,320 shichen/shi/dashi = 129,600 fen/years/yuan
- 1 dayuan = $129{,}600^2$ years = 16.8 billion years

Table 5.3 lists the quantitative relations of the numeral coefficients in the twelve and thirty categories. The five pillars of all the tiers are expressed in a scale-free cycle by an identical set of periods, S_1, S_2, S_3, S_4, and S_5, with

$$S_1 = 12S_2 = 360S_3 = 4{,}320S_4 = 129{,}600S_5$$

where the five periods, S_1, S_2, S_3, S_4, and S_5, represent, respectively:

- year, month, ri, shichen, and fen in tier 1
- yuan, hui, yun, shi, and year in tier 2
- dayuan, dahui, dayun, dashi, and yuan in tier 3

TABLE 5.3. CALENDRICAL COEFFICIENTS

Category	1	12	12 × 30	12 × 30 × 12	12 × 30 × 12 × 30
S_1	1	12	360	4,320	129,600
S_2	1 / 12	1	30	360	10,800
S_3	1 / 360	1 / 30	1	12	360
S_4	1 / 4,320	1 / 360	1 / 12	1	30
S_5	1 / 129,600	1 / 10,800	1 / 360	1 / 30	1
Period	S_1	S_2	S_3	S_4	S_5

Scale-Free Quantification

In the S_1 cycle, the twelve powers energize the twelve S_2 periods. The periods are further quantified into shorter pillar periods such as S_3. Starting on the next page, table 5.4 gives a complete scale-free quantification. For convenience in recognizing the three tiers, tiers 1 and 3 are in red, in units of fen and yuan, respectively; tier 2 is in blue, in the units of the regular Gregorian year.

Tier 3 has a total of 129,600-yuan cycles. Because yuan is also a cycle of 129,600 years, this Tier thus has 129,600 × 129,600 = 16.8 billion years, designated as a celestial cycle. These yuan cycles form a giant numero-cosmological calendrical chronology of 12 dahui, 360 dayun, 4,320 dashi. Such a yuan cycle has its ruling hexagrams that can be retrieved, where the ruling hexagram of any yuan determines any of the years at tier 2, and the ruling hexagram of any year determines any of the fen at tier 3. Upon determining the ruling hexagrams of any tier, the cosmo-ecological status of each unit among the 129,600 ones in the tier can be ascertained by checking the I Ching verses. This is a procedure holographically extending Shao Yong's scheme to a higher tier, which is then generalized into a scale-free calendrical catalog (Ma and Zeng 2024).

TABLE 5.4. SCALE-FREE QUANTIFICATION OF SI

(Tiers 1 and 3 in Red; Tier 2 in blue)

45 Li	S_2 P2	post-zi, Winter Solstice: 270° – 285°			pre-chou, Less Cold: 285° – 300°		10,800
		13 Lvv (5400): 1 – 5,400 / 67017 BCE – 61618 BCE			61 Dayou (5400): 5,401 – 10,800 / 61617 BCE – 56218 BCE		
	$6S_3$ ×5	32 Fu (2160)	33 Yi (2160)	34 Zhun (2160)	35 Yii (2160)	36 Zhen (2160)	
		1 – 2,160 67017 BCE – 64858 BCE	2,161 – 4,320 64857 BCE – 62698 BCE	4,321 – 6,480 62697 BCE – 60538 BCE	6,481 – 8,640 60537 BCE – 58378 BCE	8,641 – 10,800 58377 BCE – 56218 BCE	
	S_2 P3	post-chou, Great Cold: 300° – 315°			pre-yin, Spring Beginning: 315° – 330°		21,600
		37 Shihe (5400): 10,801 – 16,200 / 56217 BCE – 50818 BCE			41 Bii (5400): 16,201 – 21,600 / 50817 BCE – 45418 BCE		
	$6S_3$ ×5	37 Shihe (2160)	38 Sui (2160)	39 Wuwang (2160)	40 Mingyi (2160)	41 Bii (2160)	
		10,801 – 12,960 56217 BCE – 54058 BCE	12,961 – 15,120 54057 BCE – 51898 BCE	15,121 – 17,280 51897 BCE – 49738 BCE	17,281 – 19,440 49737 BCE – 47578 BCE	19,441 – 21,600 47577 BCE – 45418 BCE	
	S_2 P4	post-yin, Rain Water: 330° – 345°			pre-mao, Insect-Waking: 345° – 360°(0°)		32,400
		47 Tongren (5400): 21,601 – 27,000 / 45417 BCE – 40018 BCE			44 Feng (5400): 27,001 – 32,400 / 40017 BCE – 34618 BCE		
	$6S_3$ ×5	42 Jiji (2160)	43 Jiaren (2160)	44 Feng (2160)	46 Ge (2160)	47 Tongren (2160)	
		21,601 – 23,760 45417 BCE – 43258 BCE	23,761 – 25,920 43257 BCE – 41098 BCE	25,921 – 28,080 41097 BCE – 38938 BCE	28,081 – 30,240 38937 BCE – 36778 BCE	30,241 – 32,400 36777 BCE – 34618 BCE	
63 Qian	S_2 P5	post-mao, Spring Equinox: 0° (360°) – 15°			pre-chen, Pure Brightness: 15° – 30°		43,200
		31 Gou (5400): 32,401 – 37,800 / 34617 BCE – 29218 BCE			47 Tongren (5400): 37,801 – 43,200 / 29217 BCE – 23818 BCE		
	$6S_3$ ×5	48 Lin (2160)	49 Sun (2160)	50 Jie (2160)	51 Zhongfu (2160)	52 Guimei (2160)	
		32,401 – 34,560 34617 BCE – 32458 BCE	34,561 – 36,720 32457 BCE – 30298 BCE	36,721 – 38,880 30297 BCE – 28138 BCE	38,881 – 41,040 28137 BCE – 25978 BCE	41,041 – 43,200 25977 BCE – 23818 BCE	
	S_2 P6	post-chen, Grain Rain: 30° – 45°			pre-si, Summer Beginning: 45° – 60°		54,000
		55 Lv (5400): 43,201 – 48,600 / 23817 BCE – 18418 BCE			59 Xiaoxu (5400): 48,601 – 54,000 / 18417 BCE – 13018 BCE		
	$6S_3$ ×5	53 Kui (2160)	54 Dui (2160)	55 Lv (2160)	56 Tai (2160)	57 Daxu (2160)	
		43,201 – 45,360 23817 BCE – 21658 BCE	45,361 – 47,520 21657 BCE – 19498 BCE	47,521 – 49,680 19497 BCE – 17338 BCE	49,681 – 51,840 17337 BCE – 15178 BCE	51,841 – 54,000 15177 BCE – 13018 BCE	
	S_2 P7	post-si, Less Fullness: 60° – 75°			pre-wu, Grain in Ear: 75° – 90°		64,800
		61 Dayou (5400): 54,001 – 59,400 / 13017 BCE – 7618 BCE			62 Guai (5400): 59,401 – 64,800 / 7617 BCE – 2218 BCE		
	$6S_3$ ×5	58 Xv (2160)	59 Xiaoxv (2160)	60 Dazhuang (2160)	61 Dayou (2160)	62 Guai (2160)	
		54,001 – 56,160 13017 BCE – 10858 BCE	56,161 – 58,320 10857 BCE – 8698 BCE	58,321 – 60,480 8697 BCE – 6538 BCE	60,481 – 62,640 6537 BCE – 4378 BCE	62,641 – 64,800 4377 BCE – 2218 BCE	

TABLE 5.4. SCALE-FREE QUANTIFICATION OF SI (*cont.*)

(Tiers 1 and 3 in Red; Tier 2 in blue)

18 Kan	S_2 P8	post-wu, Summer Solstice: 90° – 105°			pre-wei, Less Heat: 105° – 120°		75,600
		50 Jie (5400): 64,801 – 70,200 / 2217 BCE – 3183			2 Bi (5400): 70,201 – 75,600 / 3184 – 8583		
	$6S_3$ ×5	31 Gou (2160)	30 Daguo (2160)	29 Ding (2160)	28 Heng (2160)	27 Xun (2160)	
		64,801 – 66,960 2217 BCE – 58 BCE	66,961 – 69,120 57 BCE – 2103	69,121 – 71,280 2104 – 4263	71,281 – 73,440 4264 – 6423	73,441 – 75,600 6424 – 8583	
	S_2 P9	post-wei, Great Heat: 120° – 135°			pre-shen, Autumn Beginning: 135° – 150°		86,400
		26 Jing (5400): 75,601 – 81,000 / 8584 – 13983			22 Kunn (5400): 81,001 – 86,400 / 13984 – 19383		
	$6S_3$ ×5	26 Jing (2160)	25 Gu (2160)	24 Sheng (2160)	23 Song (2160)	22 Kunn (2160)	
		75,601 – 77,760 8584 – 10743	77,761 – 79,920 10744 – 12903	79,921 – 82,080 12904 – 15063	82,081 – 84,240 15064 – 17223	84,241 – 86,400 17224 – 19383	
	S_2 P10	post-shen, Heat End: 150° – 165°			pre-you, White Dew: 165° – 180°		97,200
		16 Shi (5400): 86,401 – 91,800 / 19384 – 24783			19 Huan (5400): 91801 – 97,200 / 24784 – 30183		
	$6S_3$ ×5	21 Weiji (2160)	20 Xie (2160)	19 Huan (2160)	17 Meng (2160)	16 Shi (2160)	
		86,401 – 88,560 19384 – 21543	88,561 – 90,720 21544 – 23703	90,721 – 92,880 23704 – 25863	92,881 – 95,040 25864 – 28023	95,041 – 97,200 28024 – 30183	
0 Kun	S_2 P11	post-you, Autumn Equinox: 180° – 195°			pre-xv, Cold Dew: 195° – 210°		10,8000
		32 Fu (5400): 97,201 – 102,600 / 30184 – 35583			16 Shi (5400): 102,601 – 108,000 / 35584 – 40983		
	$6S_3$ ×5	15 Dun (2160)	14 Xian (2160)	13 Lüü (2160)	12 Xiaoguo (2160)	11 Jiann (2160)	
		97,201 – 99,360 30184 – 32343	99,361 – 101,520 32344 – 34503	101,521 – 103,680 34504 – 36663	103,681 – 105,840 36664 – 38823	105,841 – 108,000 38824 – 40983	
	S_2 P12	post-xv, Frost-Descending: 210° – 225°			pre-hai, Winter Beginning: 225° – 240°		118,800
		8 Qiann (5400): 108,001 – 113,400 / 40984 – 46383			4 Yv (5400): 113,401 – 118,800 / 46384 – 51783		
	$6S_3$ ×5	10 Jian (2160)	9 Gen (2160)	8 Qiann (2160)	7 Pi (2160)	6 Cui (2160)	
		108,001 – 110,160 40984 – 43143	110,161 – 112,320 43144 – 45303	112,321 – 114,480 45304 – 47463	114,481 – 116,640 47464 – 49623	116,641 – 118,800 49624 – 51783	
	S_2 P1	post-hai, Less Snow: 240° – 255°			pre-zi, Great Snow: 255° – 270°		129,600
		2 Bi (5400): 118,801 – 124,200 / 51784 – 57183			1 Bo (5400): 124,201 – 129,600 / 57184 – 62583		
	$6S_3$ ×5	5 Jin (2160)	4 Yv (2160)	3 Guan (2160)	2 Bi (2160)	1 Bo (2160)	
		118,801 – 120,960 51784 – 53943	120,961 – 123,120 53944 – 56103	123,121 – 125,280 56104 – 58263	125,281 – 127,440 58264 – 60423	127,441 – 129,600 60424 – 62583	

To give a sense of the catalog, appendix B shows 325 continuous yuan cycles of 106,312 to 106,636, corresponding to the Planck observational error of the age of our universe. These cycles are in the contracting phase, 2 Kun, in figure 3.7. They are powered by P11 in dahui S_2, falling in *you* of the twelve terrestrial branches, and governed by the 11 Jiann ䷴. Its mandate is as follows:

> The image of the Hexagram symbolizes the Progressive Advance; the Hexagram suggests to us the marriage of a young lady, and the good fortune (attending it); There will be an advantage in being firm and correct; The advance indicated by Jiann is (like) the marrying of a young lady who is attended by good fortune; (The lines) as they advance to get into their correct places; This indicates the achievements of successful progress; The advance is made according to correctness; (The subject of the hexagram) might rectify his country; Among the places (of the hexagram) is the undivided line in the center; In (the attributes of) restfulness and flexible penetration we have (the assurance of) an (onward) movement that is inexhaustible. (Sung 1973, pp. 223–24, c.f. Legge 1963, pp. 178–79, 257, 333)

For interested readers, more definite explanations and/or predictions are available upon request for any related unit of any tiers.

Power Distributions and Redistributions

Let us take tier 2 (in table 5.4) as an example to illustrate the power distributions and redistributions. The four principal hexagrams (zhenggua)—45 Li ䷝, 63 Qian ䷀, 18 Kan ䷜, and 0 Kun ䷁—govern the four seasons, 32,400 years for each (Andreeva and Steavu 2016, p. 193). They are also responsible for the intercalations of timekeeping, whenever necessary (e.g., Shao 2003, pp. 3–4). The reason of the intercalations is that a solar year does not last exactly 12 lunar months. It is rather about 12.37 months. It is thus necessary to make use of these four principal hexagrams to insert intercalary days or months in a lunisolar year.

TABLE 5.5. PRINCIPAL HEXAGRAMS, DERIVATIVES, AND WORLD-ORDERING TERMS

Hexagram	D-1	D-2	D-3	D-4	D-5	D-6
45 Li	13 Lvv Winter Solstice	61 Dayou Less Cold	37 Shihe Great Cold	41 Bii Spring Beginning	47 Tongren Rain Water	44 Feng Insect Waking
63 Qian	31 Gou Spring Equinox	47 Tongren Pure Brightness	55 Lv Grain Rain	59 Xiaoxu Summer Beginning	61 Dayou Less Fullness	62 Guai Grain in Ear
18 Kan	50 Jie Summer Solstice	2 Bi Less Heat	26 Jing Great Heat	22 Kunn Autumn Beginning	16 Shi Heat End	19 Huan White Dew
0 Kun	32 Fu Autumn Equinox	16 Shi Cold Dew	8 Qiann Frost Descending	4 Yv Winter Beginning	2 Bi Less Snow	1 Bo Great Snow

Only in this way can the lunisolar cycle be synchronized with the seasons or the Moon phases in any given year.

Every principal hexagram has 6 derivatives, written in order from D-1 to D-6. These derivatives are available by changing the polarities of the hexagram's individual lines following a specific sequence. For instance, the first derivative of the principal hexagram, 63 Qian ䷀, is D-1, 31 Gou ䷫. This derivative is obtained by changing the bottom line of 63 from its original polarity — to --. Similarly, the second derivative D-2 is 47 Tongren ䷌. This derivative is obtained by changing the second line, counted from the bottom, of 63 from its original polarity — to -- . The same procedure is used to obtain D-3 to D-6 by changing the polarities of the third, fourth, fifth, and the top line of 63.

As listed in table 5.5, the 4 principal hexagrams produce 24 derivatives in total. These derivatives dominate the set of 24 world-ordering terms, identical to 24 solar terms but in a higher tier, including 12 nodal ones and 12 medial ones. Due to the holography, these terms have the same names as those assigned to the 24 solar terms in the lower Tier 1. Every derivative stands for 5,400 years. The 24 derivatives correlate with 60 active hexagrams. These 60 hexagrams are given in table 5.4 continuously, starting from 32 Fu and ending at 1 Bo. There are altogether

3 × 4 = 12 such lines with 5 elements of each, corresponding to the 60 active hexagrams. They are distributed evenly in the circular array of the square-circular diagram of figure 3.3 (p. 57). They follow in a clockwise direction but exclude the 4 principal hexagrams. Every active hexagram stands for 2,160 years. All the 60 active hexagrams represent a complete Cycle of 60 × 2,160 = 129,600 years, lasting the historical period from 67017 BCE to 62583.

Shao Yong's scheme provides the following results:

- Life on Earth began between 40,017 BCE and 34,618 BCE. It happened during the first season, 45 Li, and fell within the pre-mao after the post-*yin* of the third epoch. More precisely, it occurred in the term of Insect Waking (IW) dominated by the derivative, D-6, of Li, 44 Feng.
- Life will come to an end between 40,984 and 46,383. It will happen during the last season, 0 Kun, and fall within the eleventh epoch, Xv. More precisely, it will occur in the term of Frost Descending (FD) dominated by the derivative, D-3, of Kun, 8 Qian.
- Year 2024 was located in the third season, 18 Kan, within the seventh epoch, Wu. Precisely, the current term was Summer Solstice (SS) dominated by the derivative, D-1, of Kan, 50 Jie, ranging from 2217 BCE to 3183.

During this term, humans are governed by one of the 60 active hexagrams, 30 Daguo, which presides over the 2,160-year era from 57 BCE to 2103. This hexagram's name can be translated as "a beam that is weak," and "there will be an advantage in moving (under its conditions) in any direction whatever; there will be a success" (Sung 1973, p. 123).

Quantified Derivatives and Subderivatives

After the power distributions and redistributions, the active hexagram (say, 30 Daguo) is known to preside over a 2,160-year era. All the annual hexagrams of the 2,160 years are then determined year after year. The determination follows a concisely ordered numerological pattern around the square-circular diagram of figure 3.3. The heavenly mandate of the annual cycle can thus be available from the interpretations of the I Ching as well as its Ten Commentaries. We can take the ongoing Revolution (yun) as an example.

Derivatives of the Active Hexagram, 30 Daguo

The active hexagram, 30 Daguo ䷛, covers the 2,160 years from 57 BCE to 2103. It has 6 derivatives; each rules over one Revolution (yun) of 360 years. The derivatives and the years are listed in the first three columns of table 5.6 (p. 128). For example, the sixth derivative on the last row is 31 Gou ䷫. It covers 360 years from 1744 to 2103. Notice that the Chinese pinyin term, *yun*, is used to replace the longer word, *revolution*. Similar changes are adopted hereafter for the tables.

Every derivative further includes 6 subderivatives by following the same procedure as before: changing the polarity of its bottom line, followed by its second one, third one, fourth one, fifth one, and the top one, successively. For example, the above-mentioned sixth derivative 31 Gou ䷫ gives its 6 subderivatives in the last row of table 5.6. There are thus 36 such subderivatives altogether, each presiding over 60 years. Note that 60 is the smallest permutation number between the ten celestial stems (starting from S1-Jia element), and the 12 terrestrial branches (starting from B1-Zi element), every hexagram of the 36 subderivatives is called a Jia-Zi hexagram. All these subderivatives are given in table 5.6, from the fourth to the ninth columns.

Subderivatives of the Sixth Derivative, 31 Gou

The 6 subderivatives of the sixth derivative, ䷫ 31 Gou, are given in table 5.6. Every subderivative is a Jia-Zi hexagram covering 60 years, conventionally known as one Jia-Zi. Note that one Jia-Zi consists of two generations (shi), of 30 years each. The fifth subderivative of the sixth derivative is ䷱ 29 Ding, obtained by changing the polarity of the fifth line of 31 Gou. It presides over two generations (60 years) from 1984 to 2043. According to the I Ching, the nature of this period is paraphrased as "the intimation of great progress and success" (Legge 1963, p. 116; Sung 1973, p. 211). Among these years we find year 2024.

TABLE 5.6. HEXAGRAM 30 DAGUO: DERIVATIVES AND SUBDERIVATIVES

Derivative		Yun	Sub-Derivative and the First-Year Jia-Zi Hexagram No. (in red)					
D-1	62 Guai	57 BCE – 303 CE	57 BCE–3 CE 30 Daguo 57 BCE: 30	4–63 46 Ge 4: 46	64–123 54 Dui 64: 54	124–183 58 Xu 124: 58	184–243 60 Dazhuang 184: 60	244–303 63 Qian 244: 31
D-2	14 Xian	304 – 663	304–363 46 Ge 304: 46	364–423 30 Daguo 364: 30	424–483 6 Cui 424: 6	484–543 10 Jian 484: 10	544–603 12 Xiaoguo 544: 12	604–663 15 Dun 604: 15
D-3	22 Kunn	664 – 1023	664–723 54 Dui 664: 54	724–783 6 Cui 724: 6	784–843 30 Daguo 784: 30	844–903 18 Kan 844: 17	904–963 20 Xie 904: 20	964–1023 23 Song 964: 23
D-4	26 Jing	1024 – 1383	1024–1083 58 Xv 1024: 58	1084–1143 10 Jian 1084: 10	1144–1203 18 Kan 1144: 17	1204–1263 30 Daguo 1204: 30	1264–1323 24 Sheng 1264: 24	1324–1383 27 Xun 1324: 27
D-5	28 Heng	1384 – 1743	1384–1443 60 Dazhuang 1384: 60	1444–1503 12 Xiaoguo 1444: 12	1504–1563 20 Xie 1504: 20	1564–1623 24 Sheng 1564: 24	1624–1683 30 Daguo 1624: 30	1684–1743 29 Ding 1684: 29
D-6	31 Gou	1744 – 2103	1744–1803 63 Qian 1744: 31	1804–1863 15 Dun 1804: 15	1864–1923 23 Song 1864: 23	1924–1983 27 Xun 1924: 27	1984–2043 29 Ding 1984: 29	2044–2103 30 Daguo 2044: 30

Note that the hexagram number of the first year in every Jia-Zi period is not always the same as that of the subderivative obtained from a derivative. This is because we have to avoid the four principal hexagrams being the active hexagram representing any year. The related four cases correspond to the shaded areas in table 5.6.

Annual Hexagrams of the Jia-Zi Hexagram, 29 Ding

The Jia-Zi hexagram 29 Ding determines the active annual hexagrams for 60 years. These hexagrams are the 60 elements that are distributed evenly moving clockwise in the circular array of the square-circular diagram (fig. 3.3, p. 57). They exclude the four principal hexagrams shaded in table 5.6.

As indicated in table 5.6, the first active annual hexagram determined by the Jia-Zi hexagram 29 Ding is "1984: 29," referring to hexagram 29 Ding for the beginning year of 1984. It controls a total of 60 active annual hexagrams which are given in table 5.7. Every hexagram provides the information for the year over which it presides. For example, the year 2017 is assigned to the hexagram 37/21, Shihe.

TABLE 5.7. ANNUAL HEXAGRAMS OF THE JIA-ZI HEXAGRAM, 29 DING (FOUR ZHENGGUA ARE SHADED)

up / down	☷	☶	☵	☴	☳	☲	☱	☰
☷	0/2 KUN	1/23 Bo 2011	2/8 Bi 2010	3/20 Guan 2009	4/16 Yv 2008	5/35 Jin 2007	6/45 Cui 2006	7/12 Pi 2005
☶	8/15 Qiann 2004	9/52 Gen 2003	10/39 Jian 2002	11/53 Jiann 2001	12/62 Xiaoguo 2000	13/56 Lüü 1999	14/31 Xian 1998	15/33 Dun 1997
☵	16/7 Shi 1996	17/4 Meng 1995	18/29 KAN	19/59 Huan 1994	20/40 Xie 1993	21/64 Weiji 1992	22/47 Kunn 1991	23/6 Song 1990
☴	24/46 Sheng 1989	25/18 Gu 1988	26/48 Jing 1987	27/57 Xun 1986	28/32 Heng 1985	29/50 Ding 1984	30/28 Daguo 2043	31/44 Gou 2042
☳	32/24 Fu 2012	33/27 Yi 2013	34/3 Zhun 2014	35/42 Yii 2015	36/51 Zhen 2016	37/21 Shihe 2017	38/17 Sui 2018	39/25 Wuwang 2019
☲	40/36 Mingyi 2020	41/22 Bii 2021	42/63 Jiji 2022	43/37 Jiaren 2023	44/55 Feng 2024	45/30 LI	46/49 Ge 2025	47/13 Tongren 2026
☱	48/19 Lin 2027	49/41 Sun 2028	50/60 Jie 2029	51/61 Zhongfu 2030	52/54 Guimei 2031	53/38 Kui 2032	54/58 Dui 2033	55/10 Lü 2034
☰	56/11 Tai 2035	57/26 Daxu 2036	58/5 Xv 2037	59/9 Xiaoxu 2038	60/34 Dazhuang 2039	61/14 Dayou 2040	62/43 Guai 2041	63/1 QIAN

This hexagram embodies the annual status as described in the I Ching: the year has a "successful progress (in the condition of things which it supposes)" and "it will be advantageous to use legal constraints" (Sung 1973, p. 95, c.f. Legge 1963, p. 101). Luckily, the year 2018 appears to be a better one owing to the annual hexagram 38/17, Sui. In this year, "[under its conditions] there will be great progress and success, but it will be advantageous to be firm and correct. There will (then) be no error" (Sung 1973, p. 79, c.f. Legge 1963, p. 93). For 2024, the active annual hexagram is 44/55, Feng, which "intimates progress and development. When a king has reached the point (which the name denotes) there is no occasion to be anxious (through fear of a change). Let him be as the Sun at noon" (Sung 1973, p. 231).

It is worth mentioning that the numeral procedure introduced above, though taking Tier 2 as an example, can be applicable for any tiers due to the holographic property as shown in table 5.4 (p. 122). In this case, the manipulation is irrelevant either for Shao Yong's mesocosmic scales of the Cycle-Epoch-Revolution-Generation system, or for the microcosmic lunisolar scales of the Year-Month-Ri-Shichen system. As a result, the numeral procedure applies for acquiring the properties of the entities in any tiers holographic to Shao Yong's scales or to the lunisolar scales. Therefore, such a procedure applies not only to predicting bodily health by examining the daily running of *qi* in the Month-Ri-Shichen-Fen system but also to the cosmological evolution of the universe by examining the astronomical observations made in the cosmic cycle of Great Cycle–Great Epoch–Great Revolution–Great Generation.

QUANTIFIED BIG-BANG UNIVERSE

In view of modern astronomy and cosmology, a macrocosmic resolution of S_2 in da-hui (about 1.4 billion years) is sufficient. It is thus unnecessary to consider shorter periods (like ice ages), where a higher mesocosmic resolution becomes mandatory such as S_4 (about 4 million years) or S_5 (about 0.13 million years).

Many different physical models have been suggested to depict the macrocosmic evolution of the universe. An extraordinary new vision was offered by Penrose, the previously introduced "conformal cyclic cosmology" (CCC) (e.g., Penrose 2010). He suggests "that the universe as a whole should be viewed as an extended conformal manifold, consisting of a (possibly infinite) succession of eons, each of which gives rise to a complete history of the expanding universe" (Penrose 2010, p. 147). He believed the universe is a giant black hole, the topology of which is determined solely by the light cones of each point in space-time, while the black hole is a dualistic entity: on the one side, it is the highest entropy source in the universe and evaporates as a result of Hawking radiation; on the other side, it is also the entropy sink where a smooth transition can be made from the end of one suggested aeon (10^9 years) to the beginning of another in a big-bang process (c.f. Cañizares 2011).

Swimmean evolutionism not only accords well with the mainly qualitative CCC theory, but also quantifies, specifies, and supplements additional physical models of the big-bang cosmology. Table 5.8 (p. 132) lists compatibilities between physical models and Swimmean evolutionism.

Furthermore, the twelve Swimmean powers quantitatively exert triplicated drives to the three energy levels in the cosmological evolution:

Top level: Twelve stages of the cosmological big-bang cycle, an aeon
Middle level: Twelve stages of the macrocosmic cycle, a Great Cycle of a total of 129,600 cycles
Bottom level: Twelve stages of the mesocosmic cycle, a Cycle of a total of 129,600 years

For lower levels of qi flows, such as in lunisolar or human cycles, the ongoing nodal-medial calendrical system and the bodily meridian system are already precise enough for practical applications.

These triplicated expressions are shown in table 5.9 (p. 133), along with the corresponding cosmological and astronomical events at the three levels of a big-bang universe. The events are collected from PBS (1997), Mastin (2019), CMS (2011), Kisak (2015), and Terzić (2017). In the Great Cycle, Power 11 drives the tenth Great Epoch, the present 106312th Yuan. It is expanded as an example to indicate the evolution of the celestial, terrestrial, and human entities. Notice that, though the same terms "Heaven" and "Earth," are used at different levels, the names have different intrinsic meanings. At the top level, for example, they indicate the "Grand Unification" and "Inflation," respectively, given modern concepts. By contrast, at the middle level, they mean "post-photon era" and "stelliferous era," respectively; and, at the bottom level of the present stage, they represent our observable universe and the homeland Earth, respectively.

TABLE 5.8. COMPATIBILITIES BETWEEN SWIMMEAN EVOLUTIONISM AND PHYSICAL MODELS

Physical model	Swimmean evolutionism
Ice Age: last glacial period ≈ 120,000 years (Clayton et al. 2006, pp.1–4)	A Cycle (yuan), period of an ice-age: 129,600 years
Oldest star, HD 140283: 14.5±0.8 Ga > 13.8 Ga (NASA 2013)	A Great Cycle (dayuan), period of a big-bang universe: 16.8 Ga; Present age: 13.8 Ga
A Platonic Year (Earth axial precession): 25,800 years (Hohenkerk et al 1992, p.99)	A phase among the 5 phases of a Cycle: 129,600 / 5 = 25,920 years
A Galactic Year: ≈220–250 Ma (Jaroszkiewicz 2016, p. 108)	A hexagram among the 60 ones of a Great Cycle: 16.8 Ga / 60 = 280 Ma
A contracting-expanding universe (Hartle & Hawking 1983)	Cyclic expansion and contraction
No contraction in an aeon (Penrose 2012)	No contraction before 7 Ga
Quantum process in both universe and brain (Hameroff and Penrose 2003, 2014)	Nature is a holographic unity of heaven, earth, and humanity
Nearly scale-invariant contraction and expansion; Inflation-expansion-contraction deceleration; Cosmic structure formed during contraction (Steinhardt and Turok 2002, 2005)	Cyclic evolution: void-expansion-contraction-void; 2.8–4.2 Ga Initiation of humanized entities; 4.2–7.0 Ga Construction of all entities; 7.0–8.4 Ga Construction climax
Expanding universe at present (Planck Collaboration 2016)	The expanding stage goes untill 7 Ga; now is a contracting universe with the contracting things observed in the lower tier appearing to fly away in the larger-scale contracting background
Horizon problem of large-scale isotropy (Hetherington 2014, p. 307)	Problem solved: isotropic in a multiverse which is holofractal and holographic, rather than in a limited observable universe
Big-crunch of collapse to happen from today (Wang et al. 2004)	Precisely, it will happen 3 Ga later from today

TABLE 5.9. TRIPLICATED SWIMMEAN POWERS IN BIG-BANG COSMOLOGY

Swimmean Power									Big-Bang Cosmology
P1	Primordial	0 – 1		$0 - 10^{-44}$ s				Nothingness	Planck
	Nebulous Void	$1-\tau$		$10^{-44} - 10^{-39}$ s				Formation of Dao-Heaven (A)	Grand Unification
P2	Great Easiness	$(1-\tau)\tau$		$10^{-39} - 10^{-34}$ s				Formation of Dao-Earth (B)	Inflation
P3	Great Origin	$(1-\tau)\tau^2$		$10^{-34} - 10^{-29}$ s				Initiation of A-B Formless Primordial Qi	1st Electroweak (EW): Inflation ends; bosons
P4	Great Inaugural	$(1-\tau)\tau^3$		$10^{-29} - 10^{-23}$ s				Construction of Form in Qi	2nd EW: Higgs field and particles
P5	Great Simplicity	$(1-\tau)\tau^4$		$10^{-23} - 10^{-18}$ s				Construction of Matter in Qi and Form	3rd EW: quark-gluon plasma
P6	1-Taiji	$(1-\tau)\tau^5$		$10^{-18} - 10^{-13}$ s				Climax: Construction of entity in Qi, Form, Matter	4th EW: end; Quark appears
P7	2-monograms	$(1-\tau)\tau^6$		$10^{-13} - 10^{-8}$ s				Qi condensation evolves initial A-B entities	Quark
P8	4-xiang	$(1-\tau)\tau^7$		10^{-8} s – 1 ms					Hadron
P9	8-gua	$(1-\tau)\tau^8$		1 ms – 4 min.					Lepton
P10	64-gua	$(1-\tau)\tau^9$		4 min. – 1 year					Big-bang nucleosynthesis
P11	64×64 Square-Circle Great Cycle (dayuan) = 129,600 Cycles (yuan)	$(1-\tau)\tau^{10}$		1 – 0.13 Ma				Termination of initial entities for next-level ones	Atom formation
		$(1-\tau)\tau^{11}$: 1 – 129,600 Cycles (yuan) = 16.8 Ga	P2	0.13 Ma – 1.4 Ga				Formation of Da-Yuan Heaven (C)	Post-photon era: 1st BH / galaxies
			P3	1.4 – 2.8 Ga				Formation of Da-Yuan Earth (D)	Stelliferous era; Proto-clusters
			P4	2.8 – 4.2 Ga				Initiation of C-D humanized entities	Milky Way/1st life-friendly star
			P5	4.2 – 5.6 Ga				Construction of all entities	Galactic disks; Superclusters
			P6	5.6 – 7.0 Ga					
			P7	7.0 – 8.4 Ga				Construction Climax	Milky Way spiral; Solar nebula
			P8	8.4 – 9.8 Ga				A-B entity-condensation evolves C-D lives	Sun / formation; Water on Earth
			P9	9.8 – 11.2 Ga					Life on Earth; Dark-energy
			P10	11.2 – 12.6 Ga					1st sexual reproduction; 10^{11} galaxies
			P11	Cycles (yuan): 12.6 – 14.0 Ga	**Prior 9,256 Cycles before Our Cycle**				≳1.0 Ma: Stelliferous Era ≳10.0 Ga: Life/Dark-energy Era ~13.4 Ga: Human Era on the Earth At present ~13.8 Ga: Expanding universe Star material recycling to new star
					106312th Cycle: (Ice-Age Cycle, IA)	P2	67017 BCE – 56218 BCE	IA-Heaven	
						P3	56217 BCE – 45418 BCE	IA-Earth	
						P4	45417 BCE – 34618 BCE	Life resurrection	
						P5	34617 BCE – 23818 BCE	Life evolution	
						P6	23817 BCE – 13018 BCE		
						P7	13017 BCE – 2218 BCE	Life climax	
						P8	2217 BCE – 8583	Civilization climax	
						P9	8584 – 19383	Degeneration	
						P10	19384 – 30183		
						P11	30184 – 40983		
						P12	40984 – 51783	Life termination	
						P1	51784 – 62583	Cycle destruction	
					Future 1,543 Yuan Cycles after Our Cycle				
			P12	14.0 – 15.4 Ga				Da-Yuan entity termination	EW
P12	Evacuation		P1	15.4 – 16.8 Ga				Da-Yuan universe destruction	

Accordingly, at different levels and in different cycles, the definition of "living beings" or "humans" should be disparate from each other owing to the entirely dispersive evolving phases of the universe. Similarly, "animals" should be disparate from "plants." Reasonably speaking, it is possible that a "plant" in another world may be equivalent to the "human" in our world; or, a flying plane in our world is recognized as a "god" by beings in another world.

Specifically, detailed as follows is the timeline of our universe in a big-bang aeon of 16.8 years (a great yuan), driven by the twelve Swimmean powers. The events are recorded from Swimme and Berry (1992, pp. 269–78), PBS (1997), Montgomery et al. (2009), CMS (2011), Swimme and Tucker (2011, pp. 119–31), BEC (2015), CSIRO (2015), and Kisak (2015).

Power 1: Downward less than 0.13 Ma / Upward greater than 13.80 Ga

In the big bang at the last singularity time, the universe of space and time begins with stupendous energy. In the initial radiation-dominated photon era before ≈0.13 Ma, the four types of interactions (i.e., gravitational, strong nuclear, weak nuclear, and electromagnetic) stabilize elementary particles before a millionth of a second and produce the primal nuclei within the first few minutes.

Power 2: Downward 0.13 Ma to 1.4 Ga / Upward 12.40 to 13.80 Ga

Within approximately 0.13 to approximately 0.38 Ma the universe is the radiation-dominated post-phase of the photon era (at approximately 0.3 Ma: 6,000 K, 0.5 eV, 10^{-10} m). Photons continue to interact with the charged protons, electrons, and nuclei. As the universe expands, the waves of radiation are stretched and diluted until today to make up the faint glow of microwaves that bathe the entire universe.

Within approximately 0.24 to approximately 0.3 Ma is the recombination/decoupling era. As the temperature of the universe falls to around

3,000 K, about the same temperature as the surface of the Sun, and its density falls, on the one hand recombination occurs when ionized hydrogen and helium atoms capture electrons to neutralize the electric charge. Neutral atoms (hydrogen, helium, and lithium) are thus formed as electrons, and atomic nuclei become bound together. With the electrons now bound to atoms, photons are no longer in thermal equilibrium with the matter. The universe becomes transparent to photons as they can now travel long distances without interacting with charged particles. Photons continue to be released in the universe, and decoupling happens since until this time they have been interacting with electrons and protons in an opaque photon-baryon fluid. These photons can now travel freely, as we see in today's cosmic background radiation, and make the earliest epoch observable. By the end of this period, the universe consists of a fog of about 75 percent hydrogen and 25 percent helium, with just traces of lithium. The 2.7 K microwave background radiation hails from this moment and thus gives us a direct picture of how matter was distributed at this early time because of the decoupling.

From approximately 0.3 Ma the matter-domination era starts and lasts until approximately 10 Ga. The energy density of matter dominates both radiation density and dark energy, resulting in a decelerated metric expansion of space. However, as the relentless expansion continues, the waves of light are stretched to lower and lower energy, while the matter decouples from radiation and travels onward mostly unaffected. Radiation is no longer energetic enough to break atoms. Electrons orbit nuclei to form atoms. In this era, galactic clouds form; the primal stars appear; the first elements are forged in the stars; the first supernovas give rise to the second- and the third-generation stars; giant galaxies evolve by swallowing smaller galaxies.

From approximately 0.4 Ma the dark-ages era (4,000K–60K) starts and ends at approximately 150 Ma. This era is from recombination to when the first atoms and the production of first stars. During this time, the only radiation emitted was the hydrogen line. Although photons exist, the universe at this time is dark, with no stars having formed.

With only very diffuse matter remaining, activity in the universe has tailed off dramatically, with deficient energy levels and huge time scales. Little of note happens, and the universe is dominated by mysterious "dark matter."

Within approximately 0.5 Ma the first atoms of hydrogen, helium, and lithium are forming; as the universe becomes transparent, hydrogen and helium come forth, and galaxies are seeded. This is the beginning of our observable universe, with particles of matter and light expanding away from a hot origin point.

From approximately 1.0 Ma the stelliferous era starts and continues until the present and may last further. It is from the first formation of Population III stars until the cessation of star formation, leaving all the stars in the form of degenerate remnants.

From approximately 10 Ma the habitable era starts and ends at approximately 17 Ma during which the chemistry of life may have begun.

From approximately 150 Ma the reionization era starts and ends at approximately 1 Ga. This is the period when the most distant astronomical objects are observable with telescopes; early stars start ionizing interstellar gas; as of 2016, the most remote galaxy observed is GN-z11, at a redshift of 11.09. The earliest "modern" Population III stars are formed in this period. The first quasars form from gravitational collapse, and the intense radiation they emit reionizes the surrounding universe, the second of two significant phase changes of hydrogen gas in the universe (the first being the recombination period). From this point on, most of the universe goes from being neutral back to being composed of ionized plasma.

From approximately 300 Ma the star-formation era starts and ends at approximately 500 Ma. Gravity amplifies slight irregularities in the density of the primordial gas and pockets of gas become more and more dense, even as the universe continues to expand rapidly. Stars ignite within these pockets, and these small, dense clouds of cosmic gas start to collapse under their gravity, becoming hot enough to trigger nuclear fusion reactions between hydrogen atoms, creating the very first stars.

The first stars are short-lived supermassive stars, a hundred or so times the mass of our Sun, known as Population III (or "metal-free") stars. Eventually Population II and then Population I stars also begin to form from the material from previous rounds of star making. More giant stars burn out quickly and explode in massive supernova events, their ashes going to form subsequent generations of stars.

From approximately 0.5 Ga the galaxy-formation and evolution era (18K–4 K) starts and ends at approximately 10 Ga. Large volumes of matter and star groups collapse to form galaxies. The earliest galaxies are pulled toward each other by gravitational attraction.

From approximately 1.0 Ga galaxies coalesce into proto-clusters.

The landmark events occurring during this stage of power 2 are:

Downward	Upward	Event
150 Ma	13.65 Ga	The oldest known star in the universe: Methuselah star
300 Ma	13.50 Ga	First black holes
500 Ma	13.30 Ga	Universe heating once again and mass star formation
600 Ma	13.20 Ga	First galaxies
630 Ma	13.17 Ga	The oldest γ-ray burst, GRB 090423
800 Ma	13.00 Ga	Limit of Hubble ultra-deep field
950 Ma	12.85 Ga	The oldest known quasar, CFHQS J2329-0301
1.10 Ga	12.70 Ga	The oldest known exoplanet, PST B1620-26
1.20 Ga	12.60 Ga	Milky Way

Power 3: Downward 1.4 to 2.8 Ga / Upward 11.00 to 12.40 Ga

Matter-domination era (starting at 0.3 Ma) continues till approximately 10 Ga. Stelliferous era (starting at approximately 1.0 Ma) continues. Galaxy-formation and evolution era (starting at approximately 0.5 Ga) continue until approximately 10 Ga. Galaxies continue to coalesce into proto-clusters.

The landmark events occurring during this stage of power 3 are:

Downward	Upward	Event
1.70 Ga	12.10 Ga	The second oldest γ-ray burst, GRB 140419A supernova
2.10 Ga	11.70 Ga	The oldest known spiral galaxy, BX442
2.30 Ga	11.50 Ga	The oldest known exoplanet in the habitable zone, Kapteyn B
2.40 Ga	11.40 Ga	Milky Way's cosmic halo
2.80 Ga	11.00 Ga	Peaked star formation rate/Universe cooling to ≈13,000 K on average

Power 4: Downward 2.8 to 4.2 Ga / Upward 9.60 to 11.00 Ga

Matter-domination era (starting at 0.3 Ma) continues until approximately 10 Ga. Stelliferous era (starting at approximately 1.0 Ma) continues. Galaxy-formation and evolution era (starting at approximately 0.5 Ga) continues until approximately 10 Ga. Galaxies continue to coalesce into proto-clusters; from approximately 3 Ga, galaxy clusters begin to form; Milky Way is formed from approximately 3.3 Ga (until approximately 6.7 Ga)

The landmark events occurring during this stage of power 4 are:

Downward	Upward	Event
3.30 Ga	10.50 Ga	Oldest stars abundant in metals
3.40 Ga	10.40 Ga	Universe habitable/first life-friendly star systems
3.90 Ga	9.90 Ga	Oldest galactic cluster, JKCS 041
4.10 Ga	9.70 Ga	Universe reaching 1/3 of its present diameter
4.20 Ga	9.60 Ga	The star formation rate of Milky Way stabilized

Power 5: Downward 4.2 to 5.6 Ga / Upward 8.20 to 9.60 Ga

Matter-domination era (starting at 0.3 Ma) continues until approximately 10 Ga. Stelliferous era (starting at approximately 1.0 Ma) continues.

Galaxy-formation and evolution era (starting at approximately 0.5 Ga) continues until approximately 10 Ga. Galaxy clusters continue to form; from approximately 5 Ga, superclusters form, and galactic disks and elliptical galaxies appear; the Milky Way continues to grow.

The landmark events occurring during this stage of power 5 are:

Downward	Upward	Event
4.45 Ga	9.35 Ga	The oldest close solar neighbor, Barnard's star
4.80 Ga	9.00 Ga	Oldest known evidence for dark energy
5.00 Ga	8.80 Ga	Milky Way's thin disk started to form
5.30 Ga	8.50 Ga	Lalande 21185
5.40 Ga	8.40 Ga	Lacaille 9352

Power 6: Downward 5.6 to 7.0 Ga / Upward 6.80 to 8.20 Ga

Matter-domination era (starting at 0.3 Ma) continues until approximately 10 Ga. Stelliferous era (starting at approximately 1.0 Ma) continues. Galaxy-formation and evolution era (starting at approximately 0.5 Ga) continues until approximately 10 Ga. Superclusters continue to form; galactic disks and elliptical galaxies continue to appear; the Milky Way continues to grow until approximately 6.7 Ga.

The landmark events occurring during this stage of power 6 are:

Downward	Upward	Event
5.80 Ga	8.00 Ga	The star formation rate of Milky Way declined
6.00 Ga	7.80 Ga	Mizar
6.30 Ga	7.50 Ga	The most powerful γ-ray burst, GRB 080319B; The fourth brightest star in the night sky, Arcturus
6.70 Ga	7.10 Ga	Universe cooling below 5 K
6.80 Ga	7.00 Ga	Growth of massive galaxies slows down
7.00 Ga	6.80 Ga	The closest star system to the Sun, Alpha Centauri

Power 7: Downward 7.0 to 8.4 Ga / Upward 5.40 to 6.80 Ga

Matter-domination era (starting at 0.3 Ma) continues until approximately 10 Ga. Stelliferous era (starting at approximately 1.0 Ma) continues. Galaxy formation and evolution era (starting at approximately 0.5 Ga) continues until approximately 10 Ga. Superclusters continue to form; galactic disks and elliptical galaxies continue to appear.

The landmark events occurring during this stage of power 7 are:

Downward	Upward	Event
7.10 Ga	6.70 Ga	Left star of Orion's belt, Alnitak
7.30 Ga	6.50 Ga	Milky Way starting to spiral
7.40 Ga	6.40 Ga	Age of an average Earth-like planet
7.80 Ga	6.00 Ga	Dark energy overtaking gravity Universe's expansion rate beginning to accelerate
8.00 Ga	5.80 Ga	Tau Ceti
8.30 Ga	5.50 Ga	Solar nebula starting to collapse Milky Way as a spiral galaxy

Power 8: Downward 8.4 to 9.8 Ga / Upward 4.00 to 5.40 Ga

Matter-domination era (starting at 0.3 Ma) continues until approximately 10 Ga. Stelliferous era (starting at approximately 1.0 Ma) continues. Galaxy-formation and evolution era (starting at approximately 0.5 Ga) continues until approximately 10 Ga. Superclusters continue to form; galactic disks and elliptical galaxies continue to appear.

Throughout the galaxies, supernova explosions spread elements, which continues to the present and will continue far into the future; the Milky Way experiences its most rapid star-formation period, and most stars form; star formation continues to the present.

At approximately 8.4 Ga, the Sun is formed within a cloud of gas in a spiral arm of the Milky Way galaxy as a late-generation star. It incorporates the debris from many generations of earlier stars. Around 8.5

to 9.0 Ga, the solar system forms as a large disk of gas and debris that swirls around this new star to give birth to planets, moons, and asteroids. Earth is the third planet out.

The landmark events occurring during this stage of power 8 are:

Downward	Upward	Event
8.60 Ga	5.20 Ga	Gliese 570
8.80 Ga	5.00 Ga	A disklike cloud floats in the Orion arm of the Milky Way galaxy
8.95 Ga	4.85 Ga	The closest star to the Sun, Proxima Centauri
9.10 Ga	4.70 Ga	The brightest supernova ever observed, SN 2005ap
9.20 Ga	4.60 Ga	Tiamat goes supernova; three supernova explosions trigger star formation in the disklike cloud
9.23 Ga	4.57 Ga	Sun is born
9.24 Ga	4.56 Ga	Sun's protoplanetary disk forms earth/planets, asteroids and comets
9.27 Ga	4.53 Ga	The Moon forms from the coalesced debris
9.35 Ga	4.45 Ga	Planets form; Earth brings forth an atmosphere, oceans, and continents
9.40 Ga	4.40 Ga	Asteroids bring water to Earth and oceans forming on Earth; first known minerals, Zircons, at Jack Hills, Australia
9.55 Ga	4.25 Ga	Right star of Orion's belt, Mintaka
9.60 Ga	4.20 Ga	Life on Earth (early estimate)
9.69 Ga	4.11 Ga	Beginning of the late heavy bombardment
9.80 Ga	4.00 Ga	Greenstone rock belt; first cell emerges

Power 9: Downward 9.8 to 11.2 Ga / Upward 2.60 to 4.00 Ga

Matter-domination era (starting at 0.3 Ma) ends at approximately 10 Ga. Stelliferous era (starting at approximately 1.0 Ma) continues. Galaxy-formation and evolution era (starting at approximately 0.5 Ga) ends at approximately 10 Ga. Supernova explosions and star formation continue.

From approximately 10.0 Ga, dark-energy dominated era (>4K) starts a little later after the formation of the solar system; matter density falls below dark-energy density (vacuum energy), and the expansion of space begins to accelerate.

At this time, the life era begins. The earliest life starts due to the cooled Earth with a refined atmosphere. Microscopic living cells, neither plants nor animals, begin to evolve and flourish in Earth's many volcanic environments.

The landmark events occurring during this stage of power 9 are:

Downward	Upward	Event
9.90 Ga	3.90 Ga	Life on Earth (late estimate); Promethio* the first prokaryotic cell, thrives by photosynthesis
10.00 Ga	3.80 Ga	End of the late heavy bombardment; oldest banded iron formation
10.20 Ga	3.60 Ga	Vaalbara supercontinent forms
10.30 Ga	3.50 Ga	The oldest evidence of life on Earth: cyanobacteria fossils, Warrawoona
10.50 Ga	3.30 Ga	Compressional tectonics
10.54 Ga	3.26 Ga	One of the most substantial impacts with a 480 km crater: Barberton event
10.60 Ga	3.20 Ga	Supercontinent forms; anoxygenic photosynthesis developed by bacteria
10.70 Ga	3.10 Ga	First land bacteria
10.80 Ga	3.00 Ga	Cyanobacteria for real photosynthesis and life produce oxygen; Moon and Mercury's geological activity is frozen
10.90 Ga	2.90 Ga	Zeta Reticuli and Kenorland supercontinent form; tides decline less than 305 m
11.00 Ga	2.80 Ga	Vaalbara supercontinent breaks up
11.09 Ga	2.71 Ga	Major komatiite eruption
11.10 Ga	2.70 Ga	First eukaryotes emerging
11.20 Ga	2.60 Ga	Oldest known giant carbonate platform

*Mention of the cells Promethio, Prospero, Kronos, and Sappho refer to Swimme and Berry (1992).

Power 10: Downward 11.2 to 12.6 Ga / Upward 1.20 to 2.60 Ga

Stelliferous era (starting at approximately 1.0 Ma) continues. Dark-energy dominated era (starting at approximately 10.0 Ga) continues. Life era (starting at approximately 10.0 Ga) continues. At approximately 12 Ga, the universe has formed some 100 billion galaxies, including our Milky Way; supernova explosions and star formation continue.

The landmark events occurring during this stage of power 10 are:

Downward	Upward	Event
11.25 Ga	2.55 Ga	Zeta Puppis forming
11.30 Ga	2.50 Ga	High oxygenation (bacterially produced oxygen accumulates in the atmosphere); movement of the tectonic plates starting; continents stabilize
11.40 Ga	2.40 Ga	Suavjarvi impact event leaves the oldest still-recognizable crater; Huronian glaciation starting, the first Snowball Earth period
11.50 Ga	2.30 Ga	Continental red beds forming; first ice ages
11.58 Ga	2.22 Ga	Atmospheric oxygen level >1%
11.60 Ga	2.20 Ga	Ozone layer forming
11.65 Ga	2.15 Ga	Huronian glaciation ending
11.70 Ga	2.10 Ga	Earliest known eukaryote fossils (acritarchs); earliest multicellular life (Francevillian group fossil)
11.78 Ga	2.02 Ga	One of the most substantial impacts, Vredefort event, leaves a 300 km crater
11.80 Ga	2.00 Ga	Prospero, a cyano-bacterium, learns to deal with oxygen and proliferates; first cells with nuclei emerge; first multicellular organisms
11.90 Ga	1.90 Ga	Atmospheric oxygen level greater than 15%; natural nuclear reactor forming in Oklo, Gabon
11.95 Ga	1.85 Ga	Sudbury impact event leaves a 250 km crater
12.00 Ga	1.80 Ga	Supercontinent Columbia forming
12.05 Ga	1.75 Ga	Oldest rocks in the Grand Canyon, Vishnu Basement Rocks
12.10 Ga	1.70 Ga	Oldest sand dune ergs
12.20 Ga	1.60 Ga	Mitochondria assimilated into eukaryotic cells
12.30 Ga	1.50 Ga	Supercontinent Columbia breaking up

Downward	Upward	Event (*cont.*)
12.35 Ga	1.45 Ga	The geologic platform covers expanding
12.40 Ga	1.40 Ga	The population of stromatolotes rapidly rising
12.43 Ga	1.37 Ga	The brightest star in Canis Minor, Procyon, coming to form
12.50 Ga	1.30 Ga	Greenville orogeny starting
12.60 Ga	1.20 Ga	Supercontinent Rodinia; Bangiomorpha pubescens (first sexually reproducing organism)

Power 11: Downward 12.6 to 14.0 Ga / Upward 1.20 Ga to 0.20 Ga later

Stelliferous era (starting at approximately 1.0 Ma) continues. Dark-energy dominated era (starting at approximately 10.0 Ga) continues. Life era (starting at approximately 10.0 Ga) continues. Supernova explosions and star formation can still be observed.

At approximately 13.4 Ga, the human era (3K) begins on Earth. The chemical processes have linked atoms to form molecules. From the dust of stars and through coded messages (DNA), humans observe the universe around them and begin to wonder from where it all came.

At the present approximately 13.8 Ga, there continue the expansion of the universe and the recycling of star materials into new stars. The present human era (starting at approximately 13.4 Ga) ends before approximately 14.0 Ga.

The landmark events occurring during this stage of power 11 are:

Downward	Upward	Event
		(I) PRIOR TO PALEOZOIC ERA
12.65 Ga	1.15 Ga	The population explosion of eukaryotes
12.70 Ga	1.10 Ga	Earliest flagellate protists
12.80 Ga	1.00 Ga	Mars's geological activity is frozen; Kronos, an early heterotrophic cell, thrives by consuming other cells; the cell Sappho reproduces through meiosis
12.95 Ga	850 Ma	The population of stromatolites declines
13.00 Ga	800 Ma	Protista population exploding

Downward	Upward	Event (*cont.*)
13.03–13.06 Ga	770–740 Ma	Kaigas glaciation of the Snowball Earth period
13.05 Ga	750 Ma	Supercontinent Rodinia breaking apart
13.08–13.14 Ga	720–660 Ma	Sturtian glaciation of the Snowball Earth period
13.10 Ga	700 Ma	Fossilized worm impressions in China; Argos, the first multicellular animal, appears
13.13–13.14 Ga	650–635 Ma	Marinoan glaciation of the Snowball Earth period
13.22 Ga	575 Ma	Avalon explosion (Ediacaran fauna emerging)
		(2) PALEOZOIC ERA
		(2.1) Cambrian
13.25 Ga	550 Ma	The invention of the shell by trilobites, clams, and snails
13.26 Ga	542 Ma	The Mesocosm: jellies, sea pens, flatworms
13.26–13.28 Ga	540–520 Ma	Cambrian explosion (Exponential diversification of life / most major animal phyla appear)
13.27 Ga	525 Ma	First trilobites
		(2.2) Ordovician
13.29 Ga	510 Ma	Vertebrate animals
13.31 Ga	488 Ma	Cambrian extinctions: 80–90% of species eliminated
13.32	480 Ma	A single landmass joined from supercontinent Gondwana, South America, Africa, Antarctica, & Madagascar
13.35 Ga	450 Ma	Plants and arthropods colonizing the land
	445 Ma	An Ordovician extinction event (60% of all species go extinct)
13.36 Ga	440 Ma	Ordovician catastrophe
		(2.3) Silurian
13.38 Ga	425 Ma	Jawed fish appear; life moves ashore
13.38 Ga	420 Ma	First air-breathing animals
13.39 Ga	415 Ma	Development of the fin
		(2.4) Devonian
13.40 Ga	395 Ma	First tetrapods; insects
13.42 Ga	380 Ma	First tree-like plants; lungs appear in fish

Downward	Upward	Event *(cont.)*
13.43 Ga	370 Ma	Late Devonian extinction (70% of all species go extinct); Devonian catastrophe; the invention of the wood cell by the lycopods; the first trees; vertebrates go ashore; amphibians
		(2.5) Carboniferous
13.45 Ga	350 Ma	Land-worthy seeds by the conifers
13.47 Ga	330 Ma	First amniotes; insects develop wings
13.48 Ga	320 Ma	First reptiles
13.49 Ga	313 Ma	Reptiles show up for land-worthy eggs
	305 Ma	Atmospheric oxygen level peaking at 35%; Arthropods growing into unprecedented sizes
13.50 Ga	300 Ma	Last supercontinent, Pangaea, forms
		(2.6) Permian
13.54 Ga	256 Ma	Therapsids, warm-blooded reptiles
13.55 Ga	251 Ma	A Permian extinction event (the largest extinction: 90~96% of all species go extinct)
	250 Ma	The brightest star in the night sky, Sirius, forms
13.56 Ga	245 Ma	Permian extinction: 75-95% of all species are eliminated
		(3) MESOZOIC ERA
		(3.1) Triassic
13.57 Ga	235 Ma	Dinosaurs appear; flowers spread
	231 Ma	First dinosaurs
13.58 Ga	225 Ma	First mammals
	220 Ma	Pangaea (supercontinent) is complete and appears; all continents joined as a single supercontinent
	216 Ma	First mammals
13.59 Ga	210 Ma	Birth of the Atlantic Ocean; the breakup of Pangaea
13.60 Ga	201 Ma	A Triassic extinction event (75% of all species go extinct)
13.61 Ga	190 Ma	First giant sauropods
13.62 Ga	180 Ma	Pangaea splitting into Laurasia and Gondwana
	176 Ma	First stegosaurs

Downward	Upward	Event *(cont.)*
13.64 Ga	155 Ma	First birds
		(3.2) Jurassic
13.65 Ga	150 Ma	Birds increase in diversity and abundance
	145 Ma	Madagascar splitting from Africa
13.67 Ga	130 Ma	First flowering plants; Laurasia and Gondwana drift apart
		(3.3) Cretaceous
13.68 Ga	125 Ma	Marsupial mammals
13.69 Ga	114 Ma	Placental mammals
	106 Ma	Largest of all carnivorous dinosaurs, Spinosaurus, evolves
13.70 Ga	100 Ma	First bees
13.71 Ga	90 Ma	Indian subcontinent splitting from Gondwana
13.72 Ga	80 Ma	Australia splitting from Antarctica
13.73 Ga	70 Ma	Primates on the scene
	68 Ma	Tyrannosaurus rex evolves
	66 Ma	A cretaceous extinction event (75% of all species, including all non-avian dinosaurs, go extinct)
13.74 Ga	65 Ma	Cretaceous extinction
	60 Ma	Primates begin to evolve
		(4) CENOZOIC ERA
		(4.1) Paleocene
13.75 Ga	55 Ma	First modern birds; rodents, bats, early whales/horses, pre-monkeys
	50 Ma	Indian subcontinent's collision with Asia: Tibet/Himalayas rise
	47 Ma	Whales returning to the water
		(4.2) Eocene
13.76 Ga	40 Ma	Antarctic ice cap grows; various orders of mammals complete
	37 Ma	Cosmic impact; Eocene catastrophe
		(4.3) Oligocene
13.76 Ga	36 Ma	Monkeys

Downward	Upward	Event *(cont.)*
13.77 Ga	35 Ma	Grasslands widespread; early cats and dogs
	30 Ma	South America separating from Antarctica; first apes
	26 Ma	First elephants
13.78 Ga	25 Ma	Whales become the largest marine animals of all time; carnivores take to the sea and become seals
		(4.4) Miocene
13.78 Ga	24 Ma	Grass spreads across land
	20 Ma	Monkeys and apes split
	19 Ma	Early antelopes
	18 Ma	Separation of great and lesser apes
	15 Ma	First bovids
13.79 Ga	12 Ma	Gibbons
	11 Ma	First large horses
	10 Ma	Orangutans
	9.00 Ma	Gorillas
	8.00 Ma	The brightest star in Orion, Rigel, forms; modern cats
	7.50 Ma	2nd brightest star in Orion, Betelgeuse, forms
	7.00 Ma	Elephants
	6.00 Ma	Last common ancestor of humans and chimpanzees; modern dogs
	5.96-5.33 Ma	Messinian Salinity Crisis (Strait of Gibraltar closes tight; the Mediterranean Sea dries out)
		(4.5) Pliocene
13.79 Ga	5.00 Ma	Chimpanzee, hominids; *Australopithecus afarensis* on the scene
13.80 Ga	4.50 Ma	Modern camels, bears, and pigs
	4.00 Ma	*Australopithecus* evolves; Baboons
	3.70 Ma	Modern horses
	3.50 Ma	Early cattle
	3.30 Ma	Most recent ice ages (a series of glacial and interglacial periods) begin

Downward	Upward	Event (*cont.*)
	3.00 Ma	Great American interchange (Isthmus of Panama rises; North and South America join; extensive fauna exchange follows)
		Lower Paleolithic Period
13.80 Ga	2.60 Ma	First humans; *Homo habilis*
	2.50 Ma	Most recent Ice Age (Last Glacial Maximum) begins
	2.20 Ma	First members of the genus Homo with tool-making abilities appear
	1.80 Ma	Modern big cats, bison, sheep, wild hogs
		(4.6) Pleistocene
13.80 Ga	1.50 Ma	First controlled use of fire by *Homo erectus*; hunters
	1.00 Ma	Mammalian peak
	780 ka	Last reversal of the Earth's magnetic field
	730 ka	Cosmic impact; Pleistocene catastrophe
	700 ka	Brown bears
	650 ka	Wolves
	640 ka	Yellowstone Caldera erupting
	500 ka	Earliest human engravings: zig-zag patterns on shells in Java; llamas; clothing, shelter, fire, hand axes
	300 ka	Earliest instances of burial; archaic *Homo sapiens*
	250 ka	Neanderthals evolve
	200 ka	Cave bears, goats, modern cattle; the earliest evidence of human art in the caves of South Africa
	150 ka	Wooly mammoth
	195 ka	Anatomically modern humans appear in Africa
	120 ka	Wildcats
		Middle Paleolithic Period
13.80 Ga	100 ka	Humans migrate out of Africa; ritual burials
	72 ka	Polar bears
	70 ka	Toba super-volcanic eruption; human population falling to ~10,000

Downward	Upward	Event (*cont.*)
		Upper Paleolithic Period
13.80 Ga	40 ka	Modern *Homo sapiens*; language; occupation of Australia
	38 ka	First domesticated dogs; Neanderthals go extinct
	35 ka	Occupation of the Americas
		Aurignacian Period
13.80 Ga	32 ka	Musical instruments
		Gravettian Period
13.80 Ga	20 ka	Spears, bows, and arrows
		Magdalenian Period
13.80 Ga	18 ka	Cave paintings in southern Europe
		Neolithic Period
13.80 Ga	12.0 ka	Dogs tamed
	10.7 ka	Sheep and goats tamed in the Middle East
	10.6 ka	Settlements in the Middle East: wheat and barley cultivated
	10.5 ka	Pre-historical sects and cults formed in Southeast Asia
	10.0 ka	Dogs tamed in North America
	9.0 ka	Settlements in Southeast Asia: painted pottery culture; rice gardeners; water buffalo, pigs, and chickens tamed
	8.8 ka	Settlements in the Middle East: cattle tamed
	8.5 ka	Settlements in the Americas: corn, squash, peppers, and beans; Settlements in the Middle East: weaving
	8.0 ka	Settlements in the Middle East: irrigation; the population of Jericho is 2,000
	7.5 ka	Hassuna culture; millet farmers in North China
	7.0 ka	Catal Hiiyuk population is 5,000
	6.4 ka	Horses tamed in Eastern Europe
	6.0 ka	Human civilization: span of recorded history
	5.3 ka	Pottery in the Andes
	5.0 ka	Early European settlements: gourds, squash, cotton, amaranth, and quinoa in the Andes; Settlements in the Middle East: camels and donkeys tamed; Settlements in India: elephants tamed
	4.5 ka	Peanuts in the Andes

Downward	Upward	Event *(cont.)*
	3.5 ka	World population is 5-10 million people
	Civilization Period (ancient, medieval, and modern)	
	Future events	
13.80 Ga	13 ka later	Earth's axial tilt reversal to reverse seasons
	50 ka later	Niagara Falls to erode away
	200 ka later	99% of human structures to be destroyed without a trace
14.0 Ga	230 Ma later	Planetary orbits impossible to be predicted

Power 12: Downward 14.0 to 15.4 Ga / Upward 0.20 to 1.60 Ga later

Stelliferous era (starting at approximately 1.0 Ma) may continue. Dark-energy dominated era (starting at approximately 10.0 Ga) continues. Life era (starting at approximately 10.0 Ga) begins to fade (until 16.6 Ga)

The landmark event occurring during this stage of power 12 is:

Downward	Upward	Event
14.6 Ga	800 Ma later	Multicellular life impossible

Power 1: Downward 15.4 to approximately 16.8 Ga / Upward 1.60 to approximately 3.0 Ga later

Dark-energy dominated era (starting at ≈10.0 Ga) may continue. Life era (starting at ≈10.0 Ga) ends before ≈16.8 Ga.

The landmark events occurring during this stage of power 1 are:

Downward	Upward	Event
16.6 Ga	2.8 Ga later	The lower bound for the next singularity time (Jiménez et al. 2016)
16.8 Ga	3 Ga later	Remaining lifeforms and remnant matters of the present universe disappear, and the universe may enter the next eon cycle (Penrose 2012). However, the process may last to a time longer than 3 Ga later (e.g., 8 Ga later in BEC 2015) or even further.

A TORUS UNIVERSE: MODEL OF CYCLIC BLACK-HOLE COSMOLOGY

The evolving pattern of natural things at different scales fits with a scale-free cycle as shown in figure 3.7 (p. 77). That is to say, all the entities own the same holographic property along the evolving direction of respective tiers in the universe. The general holographic pattern fits with the process from the void to returning to the void, and keeps going repeatedly from the past to the future. We rewrite the features of the everlasting cyclic evolution as follows:

- first emerging from the void resultant of the prior cycle;
- then growing and expanding to the maximum;
- afterward experiencing a contracting and decaying process;
- finally returning to the void again for the beginning of the next cycle.

In modern cosmology, the big-bang model is considered the most successful explanation for the evolution of the universe. Numerous studies have demonstrated its consistency with Einstein's theory of general relativity, the Hubble expansion law, galactic distribution, and cosmic microwave background radiation. Nevertheless, its validity continues to attract more and more compelling studies as both theoretical and observational research progress (Van Flandern 2002). Fundamental questions include, but are not limited to:

- Is the universe flat?
- Is there dark matter and dark energy?
- How did the big bang begin to evolve from a Planck-scale singularity?
- What is the conformity between the initial quantum-gravity singularity and the present materialized universe?
- Is the universe alive only for one cycle, or is it evolving cyclically as predicted by the holographic pattern of figure 3.7?

Modeling

Einstein's theory of relativity supersedes the two-hundred-year-old Newtonian gravitational cosmology. It intimately connects with the geometry of space and time. It proves that matter of any mass distorts the geometry; and thus, a curved geometry is the expression of gravity. The distorted one can be described by Einstein's 3-D hyper-cylinder at a constant time, t, and more accurately, by de Sitter's 4-D hyper-hyperboloid where t is a variable (Theuns 2016).

If M is massive enough, the geometry becomes so closed that it forms a 4-D black hole (BH) which is horn-torus shaped. The topology is obtained when observed in our conventional 3-D space, with two geometric dimensions suppressed (*Mathematics and Nature* 2017). Figure 5.2 illustrates the coordinates of Einstein's and De Sitter's models, and the horn-torus geometry, where two dimensions are suppressed, by updating "Lissajous figure on horn torus."

Thus, the universe is a 3-D surface membrane of the horn-torus,the

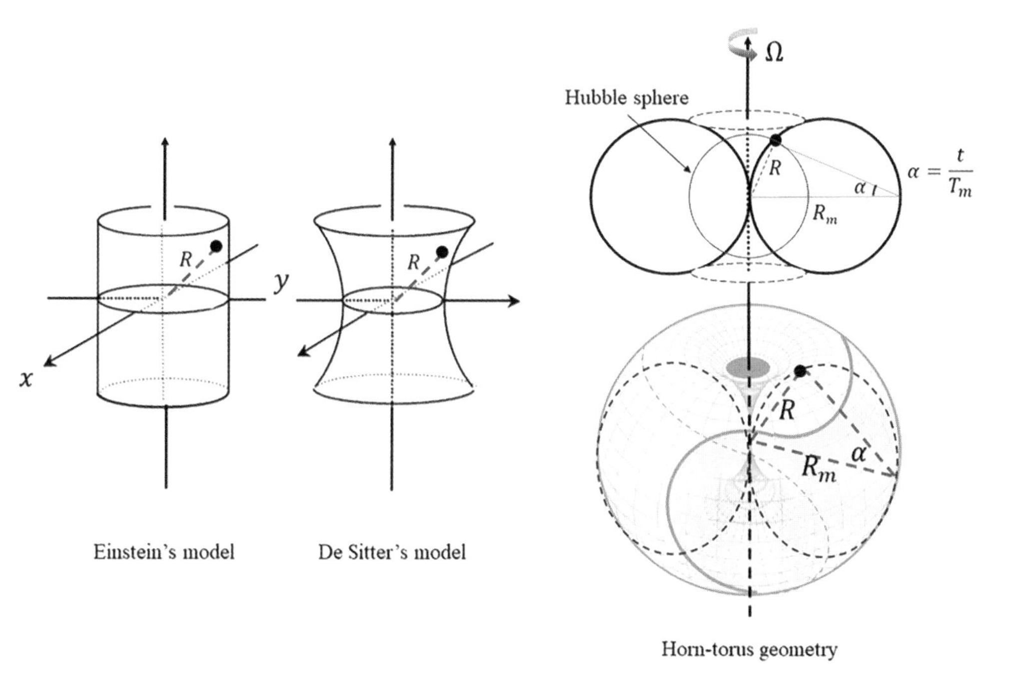

Figure 5.2. Coordinates and horn-torus geometry

center of which is the quantum big-bang singularity (to be proved below). Its evolution is represented by the axisymmetric event-horizon radius, R, of a BH, as viewed by a distant observer in our flat space (Frolov and Novikov 2012, p. 48; Halpern and Tomasello 2016; Misner et al. 1973, p. 667). If the BH rotates (denoted by Ω), the universe evolves on the surface from birth to birth cyclically along the trajectory of the solid blue band. If it does not rotate, the trajectory follows the circle of the blue dash line. The diameter of the circle is R_m. It is the maximal spatial size of the universe evolving from R. Note that the equation to describe the horn-torus shape is as follows:

$$R = R_m \sin \alpha \tag{1}$$

where α is the inclination angle of R relative to the equatorial plane of the horn-torus. Its volume changes from $V = 2\pi^2 R^3$ to $V_m = (4/3)\,\pi R_m{}^3$.

In contrast to the region within R where all the null geodesics of photons are warped and unable to escape to the 3-D surface membrane, the shell from R to R_p, the observable photon-horizon radius, is recognized as the region where quantum effects make either rotating Kerr black holes or nonrotating Schwarzschild ones to release Hawking radiation and yield Bekenstein-Hawking luminosity (Caughey 2016; Hawking 1975; Halyo et al. 1997; Khoo and Ong 2016, 2017; Penrose 1969). In the Schwarzschild case, $R = 2GM/c^2$, where M is its mass, $G = 6.674 \times 10^{-11}\ \mathrm{m^3 kg^{-1} s^{-2}}$ is the gravitational constant, and $c = 3 \times 10^8$ m/s is the speed of light (Abramowicz 1990; Kutner 2003, p. 148; Lestone 2007). For a spherically symmetric BH as viewed in our 3-D flat space, the BH mass density, ρ, is given by

$$\rho = \frac{3M}{4\pi R^3} \tag{2}$$

The ESA's Planck Mission obtained that the age, t_0, of our universe at present is 13.8 Ga, or, 4.35×10^{17} s; and, the Hubble constant, H_0, is 67.39 km/(s·Mpc), or, $2.19 \times 10^{-18}\ \mathrm{s^{-1}}$ (Planck Collaboration 2016). Thus, there exists a relation of $H_0 t_0 = 0.953$ which is close to 1, and the Hubble radius, $R = R_H$, is given by $R_H = c/H_0 = 1.36 \times 10^{26}$ m. Our universe is now a BH with $R_0 = R_H$, and then $R_{p0} = 3.5 \times 10^{26}$ m. It gives a mass of $M = R_0c^2/(2G) = 9.2 \times 10^{52}$ kg and a density of $\rho_0 = 8.7 \times 10^{-27}$ kg/m^3, $\cong \rho_{crit}$, together with mass-energy of $\varepsilon = Mc^2 = 8.3 \times 10^{68}$ J. The mass estimation is in the same order as those from either the Fred Hoyle formula (Kragh 1999, p. 488) or the dimensional analysis (Valev 2014). For R_{p0}, conventional ΛCDM modeling predicted a larger radius, 45.3 billion light-years or 4.3×10^{26} m (Halpern and Tomasello 2016).

The Planck force, F_{pl}, is defined by $F_{pl} = c^4/G = 1.2 \times 10^{44}$ N. For an evolving spherical universe of R_H in three dimensions, we have the ratio parameter, ω, of the cosmic equation of state:

$$\omega = \frac{p}{\rho c^2} = -\frac{\dfrac{F_{pl}}{8\pi R_H^2}}{\dfrac{M}{\frac{4}{3}\pi R_H^3}c^2} = -\frac{1}{3} \tag{3}$$

where p is the cosmic pressure and sign "–" indicates the opposite directions between p and F_{pl}. This result reproduces that formulated by Melia (2013, 2015).

Formalism

Einstein field equations describe the relationship between mass-energy and curved space-time (Einstein 1916; 1997, pp. 117–20):

$$G_{\mu\nu} + \Lambda g_{\mu\nu} = \kappa T_{\mu\nu} \text{ where } G_{\mu\nu} = R_{\mu\nu} - \frac{1}{2}Rg_{\mu\nu} \tag{4}$$

where $(\mu, \nu) = (0, 1, 2, 3)$; $G_{\mu\nu}$ is the Einstein tensor; Λ (s^{-2}) is the cosmological constant to be solved; $g_{\mu\nu}$ is the metric tensor; $\kappa = \frac{8\pi G}{c^2} = 1.866 \times 10^{-26}$ m/kg is Einstein's constant; $T_{\mu\nu}$ is the stress-energy tensor, with $T_{\mu\nu} = 0$ giving a vacuum condition; and $R_{\mu\nu}$ is the Ricci curvature tensor.

We adopt three assumptions to solve equation 4: (1) the space of constant curvatures is orthogonal to time; (2) the cosmic matter is incoherent and at relative rest; and (3) a perfect fluid satisfies equation (3), $p = \omega\rho c^2$. In this case, equation 4 is simplified to give a set of two Friedmann equations (Friedman 1922; Belenkly 2012):

$$\left(\frac{\dot{R}}{R}\right)^2 = \frac{8\pi G}{3}\rho - \frac{kc^2}{R^2} + \frac{\Lambda c^2}{3}; \quad \frac{\ddot{R}}{R} = -\frac{4\pi G}{3}\left(\rho + \frac{3p}{c^2}\right) + \frac{\Lambda c^2}{3} \tag{5}$$

where $\kappa = (-1, 0, +1)$ is the spatial curvature in any time-slice of the universe. It is dependent on whether the shape of the universe is open, flat, or closed, respectively.

Alternative to the conventional Lambda cold dark matter (ΛCDM) approach (Bull et al. 2016), equation 5 is solved here by eliminating all the density terms of baryons, dark matters, radiations (photons + neutrinos), and dark energy. At R_m, the universe evolves from the curved 4-D spacetime to the decoupled absolute 1-D time and 3-D spherical space, giving $\dot{R} = 0$ and $\kappa = 0$. Thus, using equation 2 in the first equation of equation 5 gives

$$\Lambda = -\kappa\rho = -\frac{3}{R_m^2} = -2.27 \times 10^{-52} < 0 \tag{6}$$

which expresses that a negative cosmological constant is valid in the evolution of an early universe (e.g., Farnes 2018; Prokopec 2011). Thus, it is applicable for the whole evolving cycle. Note that the value is about 5.7 times of that predicted by Lessner (2006, 2011).

Combining the two equations of equation 5 yields

$$2\frac{\ddot{R}}{R} + (1+3\omega)\left[\left(\frac{\dot{R}}{R}\right)^2 + \frac{kc^2}{R^2}\right] = (1+\omega)\Lambda c^2 \tag{7}$$

No matter what value of k is, solving equation 7 by using equation 3 yields

$$R = R_m\left|\sin\left(\frac{t}{T_m}\right)\right| \tag{8}$$

in which $T_m = R_m/c$.

Interestingly, equation 8 is identical to the horn-torus equation as shown in equation 1 if and only if $\alpha = t/T_m$. The solution is also the same as that presented with the mathematical model of a horn-torus universe (Cataldo 2016a, b).

Results

The solution of equation 8 is shown in figure 5.3. Obviously, the universe evolves cyclically in endless cycles. In every cycle, it emerges from the big-bang singularity at the end of the prior cycle, then expands to reach the maximal size, followed by a crunching process to contract to an ending point, the singularity of the next big-bang cycle. No doubt, the evolution reproduces that illustrated in figure 3.7 (p. 77), and identically signifies the scale-free holographic property of the universe: entities evolve from the void and return to the void and keep going repeatedly from the past to the future.

In addition, by examining the solution given by equation 8, we confirm that the universe originates from a quantum BH. Let $R = l_{\text{Pl}}$ and $t = t_{\text{Pl}}$, where $l_{\text{Pl}} = \sqrt{\hbar G/c^2} = 1.62 \times 10^{-35}$ m is the Planck length, where $\hbar$ is the Dirac constant, and $t_{\text{Pl}} = l_{\text{Pl}}/c = 5.39 \times 10^{-44}$ s is the Planck time. Since $\lim_{\alpha \to 0} \sin \alpha = \alpha$, equation 8 gives $l_{\text{Pl}} = ct_{\text{Pl}}$. The result satisfies the length-time relation at the Planck scales. Thus, the solution manifests that our universe comes from the quantum BH at the big-bang singularity and is still on the track of the evolution toward a final gravitational big-crunch BH.

Furthermore, the solution given by equation 8 validates our prediction of the cosmological cycle of the Great Cycle = 129,600 cycles = 16.8 Ga. Using $T_m = 16.8$ Ga and the age of the universe, $t_0 = 13.8$ Ga, equation 8 gives

$$R = 0.732R_m = 0.732cT_m = 1.16 \times 10^{26} \text{ m}$$

which is in good agreement with the value, $R_0 = 1.36 \times 10^{26}$ m, given under equation 2. That is to say, the aeon of the universe is 16.8 billion years, and at present it is 13.8 billion years old and actually in a

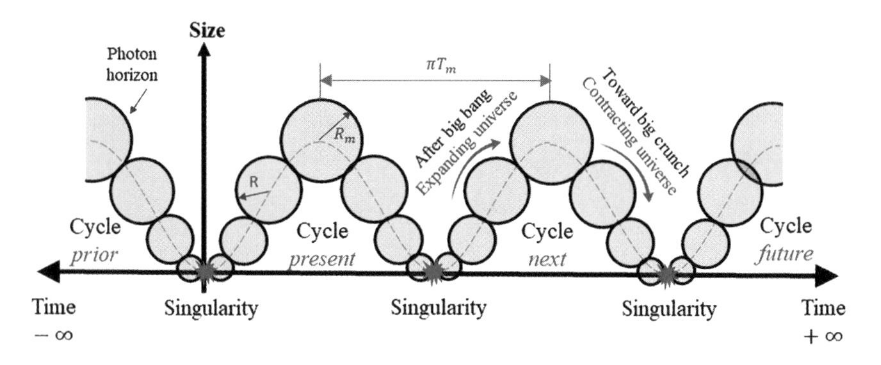

Figure 5.3. Cyclic big-bang model of a black-hole universe

contracting process as also claimed by Wetterich (2013). Accordingly, dividing T_m into a set of 12 epochs, the universe is at present at the eleventh slot, that is, powered by the eleventh Swimmean power, P11 (12.6 to approx. 14.0 Ga), of the Great Cycle shown in table 5.9. Notice that the concept of the "cycle" is different from that of an "oscillating period," $\pi T_m = 5.28$ Ga. This is a parameter to give the observable size of the universe by multiplying the speed of light, *c*. For instance, the maximal radius of the universe will be 52.8 billion light-years, while the radius of the universe at present shou be $13.8\pi = 43.4$ billion light-years, with an error of <7 percent compared to the observationally estimated 46.5 billion light-years (Bars and Terning 2009, p. 27).

Finally, equation 8 (p. 156) proves that the evolution of the universe refers to the dynamics of neither dark energy nor dark matter. In solving Friedmann equations, the only determinant is the total mass-energy of the universe. Even if the dark matter and/or energy exist, the role played by them is merely to constitute a fraction of the mass-energy, and thus partially contribute to the amount of the negative cosmological constant. I do not argue whether or not the dark matter and/or energy exist in the universe. I merely stress that they are not involved in the dynamic process of cosmological evolution as shown above. Yet we notice that some research definitely claimed their nonexistence (e.g., Mabkhout 2015, pp. 122–42; Maeder 2017; Pultarova 2017), or asserted that the dark energy and dark matter are the cosmic inertial effects. It was argued that the former comes from cosmic centrifugal rotations and the latter from a combination of the centrifugal and Coriolis forces (e.g., Seshavatharam and Lakshminarayana 2016; Zorba 2012).

VALIDATION OF PENROSE AND HAMEROFF'S UNITY OF UNIVERSE AND MAN

As discussed in chapter 1, Hawking and Penrose proposed an atheistic quantum-gravity mechanism for an evolving universe from the big-bang singularity (Hawking and Penrose 1970; Hawking 1996). Later, Penrose and Hameroff developed the previously discussed Orch-OR formulation. By means of the supposed pan-quantumistic permeability of the quantum dynamics, the authors connected the macrocosmic structures of the universe with the microcosmic neural microtubules of human brains (Hameroff and Penrose 2003, 2014).

The Orch-OR model was used to simulate brain EEG waves as measured at a rest state with closed eyes. Though the results deviated far away (see, e.g., figure 11 in Hameroff & Penrose 2014), Hawking, Penrose, and Hameroff drove the modern scientific research in a direction where it merges with ancient Eastern wisdom concerning the onesness of heaven and humanity. Unifying both the principles of general relativity and quantum theory in modern cosmology, the quantum-gravity theory has become the leading candidate nowadays in formulating the great unity of nature and man. In fact, this great unity has consistently been a mainstream philosophical theme in the history of the West, as we find it articulated in accounts by leading figures such as Kepler (McGrath 2006, p. 214), Teilhard (Mickey 2016, p. 41), and Einstein (who expressed that "a human being is a part of the whole, called by us 'Universe,' a part limited in time and space" in a letter to Marcus on February 12, 1950).

Aiming at exploring the quantitative principles of natural philosophy, this book does not focus on finding a universal mathematical model applicable to exactly simulating every cosmological phenomenon.

Instead, we pay attention to the quantifications of the evolution for all the holographic entities of the universe. This holography refers to an identical scale-free arithmetic system, applicable to any tiers of nature, and showing the authentic unity of heaven, earth, and humanity. Such a holography, in other words, recognizes that humanity can be understood as a "hominized" form of the cosmological processes energized by a series of epochal powers, in response to different cosmological circumstances at different scales (Swimme 2017a; Swimme and Busch 1990).

As one of the "hominized" products of the universe, brain consciousness should possess the trait of the pan-quantumistic permeability. Nevertheless, if the entire history of the universe is compressed into one day, human beings appear on the Earth only at the last second. Thus, the trait of the quantum effect is unable to fit 100 percent with that at the very beginning of the universe. The actual percentage can be estimated by a recently developed model of quantum-plasma brain dynamics (Q-PBD). The model solves a set of quantum-plasma Wigner-Poisson equations. In the absence of the quantum term, the PBD model effectively simulated brain EEG waves (Ma 2017; figure 5.4 below: left panel of Medical Cerebral State). When the quantum term is introduced, it is expected to play a role, at least under certain emotional states, if not under all conditions. We pay attention to this role here and present the modeling, formalism, and results as follows.

Modeling

Human consciousness resides mainly in the outer layer of the cerebrum, the cerebral cortex, where the volume density of neurons is in the order of 10^{14} neurons/m^3 on average (Teplan 2002). These neurons are interconnected with each other, each one linking with up to 10^4 other neurons, forming a highly intricate system passing signals via as many as 1,000 trillion synaptic connections (Mastin 2010).

Inside a healthy human brain, both the intracellular and extracellular spaces are concentrated with positive ions (317.5 mM) and negative ones (124.0 mM), giving their respective charge number densities of $n_+ \approx 1.9 \times 10^{26}$ m^{-3} and $n_- \approx 39\%\ n_+$ (Phillips et al. 2013, p. 685, table 17.1), with $n_+ \approx 1/1000$ of the molecular number density of water or the free electron density in copper, while the excess positive charges balance the abundant electrons coming from the macromolecules such as nucleic acids and proteins in the brain to keep the brain electrically neutral (Jibu and Yasue 1995, p. 685).

Such a high charge density in the order of 10^{26} m^{-3} makes the brain plasma distinguishable from the classical plasmas owing to the impact of the quantum effect (Manfredi 2005). This nonclassical entity requires employing the electron Wigner-Poisson equations to demonstrate the pan-quantumistic permeability in human brains through calculating the quantum effect on brain consciousness.

Formalism

The charge density in brain plasmas is so dense that the distance between particles becomes comparable to the thermal de Broglie wavelength. As a result, the quantum effect can be obtained by solving the electron Wigner-Poisson equations in the presence of an external magnetic field under the self-consistent collective electrostatic or electromagnetic conditions with some acceptable assumptions (Tyshetskiy et al. 2013, pp. 76–80). The electrons in the brain plasmas obey the perturbed equations of (e.g., Shukla and Eliasson 2010),

$$\frac{\mathrm{d}f_1}{\mathrm{d}t} = \frac{\partial f_1}{\partial t} + \mathbf{v} \bullet \nabla f_1 + \frac{\mathrm{d}\mathbf{v}}{\mathrm{d}t} \bullet \nabla_{\mathbf{v}} f_1 = -\frac{e}{m_e} \nabla \varphi_1 \bullet \nabla_{\mathbf{v}} f_0 + \frac{e\hbar^2}{24 m_e^3} \nabla^3 \varphi_1 \bullet \nabla_{\mathbf{v}}^3 f_0 \qquad (9)$$

$$\nabla^2 \varphi_1 = \frac{e}{\varepsilon_0} \int f_1 \mathrm{d}\mathbf{v} \text{ and, } \frac{\mathrm{d}\mathbf{x}}{\mathrm{d}t} = \mathbf{v},\ \frac{\mathrm{d}\mathbf{v}}{\mathrm{d}t} = -\frac{e}{m_e} \mathbf{v} \times \mathbf{B}_0 \qquad (10)$$

where f is the electron distribution function; t is time; $\mathbf{x} = \{x, y, z\}$ is position vector; $\mathbf{v} = \{\mathbf{v}_\perp, v_\parallel\}$ is velocity vector in which $\mathbf{v}_\perp = \{v_x, v_y\}$; φ is electrostatic potential; $e = 1.60 \times 10^{-19}$ C is electron charge, $m_e = 9.11 \times 10{-}^{31}$ kg is electron mass; $\varepsilon_0 = 8.85 \times 10^{-12}$ F/m is the permittivity of free space; and, ∇ and $\nabla_{\mathbf{v}}$ are the spatial gradients in the position and velocity spaces. Also, subscripts "$\perp$" and "$\parallel$" denote the components perpendicular and parallel to the magnetic field, $\mathbf{B}_0 = B_0\hat{\mathbf{e}}_z$, respectively, where $\hat{\mathbf{e}}_z$ is the unit vector along z; and subscripts "0" and "|1" denote the mean-field and perturbed components, respectively. Besides, the initial condition of $\vartheta_0 = 0$ is assumed, and the electron density takes full of n_-. Furthermore, all the positive ions are considered so heavy in mass as to be immobile but constitute the neutralizing background for the energetic electrons. Also, all the electrons are treated as if they reside well inside the extracellular space of brains.

Results

By applying the initial electron Maxwellian function to give a backward-mapping algebra,

$$f_0 = \frac{n_0}{\sqrt{\pi}} e^{-v_0^2} = \frac{n_0}{\sqrt{\pi}} e^{\varphi - v^2} \tag{11}$$

a tedious, comprehensive derivation gives

$$\frac{f_1}{f_0} = \frac{2\left[\sqrt{2}\,\pi - 2\alpha\mathbf{k}^2(3 - 2\mathbf{v}^2)\right]}{1 - \omega/(kv)} \varphi_1(\mathbf{x},t) \tag{12}$$

where $\alpha = \sqrt{2}\,\pi\eta^2/48$ is the quantum coefficient, and $n \approx 15$ is the ratio between the electron quantum energy and its thermal energy at a typical room temperature, 300K; k and ω are the electron wave number and angular frequency, respectively, while $\mathbf{k}$ is the wave number vector. For brain neuronal electrons that have the Maxwell-Boltzmann distribution, if there is no fluctuation in their thermal energy relative to the mean-field state, $\mathbf{v}^2$ turns out to be 3/2. It gives $f_1 = 0$, yielding a result that the quantum effect of α is unable to be exposed.

Alternatively, if there are fluctuations in the thermal energy, the quantum effect, R, is given by the ratio of the two terms within the square brackets, "[" and "]":

$$R \leq \frac{2\alpha k^2(3 - 2\Delta v^2)}{\sqrt{2}\,\pi} \tag{13}$$

in which Δv^2 represents the fluctuation. In the simplest one-dimensional case, $k = 1/\sqrt{\gamma_e}$ in which $\gamma_e = 3$ is the ratio of two specific heat capacities. Laboratory experiments measured the deviation of the brain temperature in a few degrees for any disorders in mental consciousness (Wang et al. 2004). Such a deviation of, say, 5° C gives rise to $\Delta v^2 = 1.67\%$, giving $R \leq 11\%$.

Consequently, we obtain following results to validate Penrose and Hameroff's model:

- As the "hominized" form of cosmological processes, brain consciousness does keep the quantum coherence, but only under abnormal situations like a sudden head trauma, mood disorder, and so forth.
- Rather than 100 percent as argued by the Orch-OR model, the quantum effect does not exist in brain consciousness at normal conscious states; and it occupies no more than 11 percent of the classical plasma effect under abnormal situations.
- Magnetic fields of the Earth do not have any influences on the quantum effects in the human brain.

Discussion

As a microcosm of the huge universe, humans rely on their mental activities to offer a profound and multidimensional lens through which we can explore the intricacies and interrelations of fields such as psychology, religion, metaphysics, and modern science. The brain's EEG waves, in particular, behave as a captivating tapestry to unfold a delicate dance of electric processes across the neural landscape. This dance finds echoes of the vast cosmic order which is mirrored within the recesses of our minds and extends its resonance into other fields.

Figure 5.4 integrates the simulated brain EEG waves with psychological forces, Buddhist consciousnesses, and Swimmean powers. The leftmost panel shows the five simulated medical cerebral states (Ma 2017), the middle left one gives the four corresponding psychological factors, the middle right panel tells the nine Buddhist mental conditions, and the rightmost panel lists the twelve Swimmean cosmogenetic inputs. This figure offers a multidimensional lens through which we can explore the intricate relationships between brain activity, psychological forces, mental perspectives, and cosmic powers. As I extend the current exploration of cosmology to include consciousness in what I hope will be a companion book, aiming to elucidate the quantum principles of reincarnation, the potential for groundbreaking insights into the nature of the mind becomes increasingly evident.

In conclusion, the exploration of brain EEG waves transcends mere neuroscience. It beckons us to embark on a holistic journey. This journey traverses the vast landscapes of the mind, and draws inspiration from key tenets of diverse fields, including ancient philosophies, metaphysics, and modern scientific contemplations. It invites an increasing number of individuals to transcend the boundaries among the self and the cosmos and collaborate for pondering the profound holographic interconnectedness of our existence.

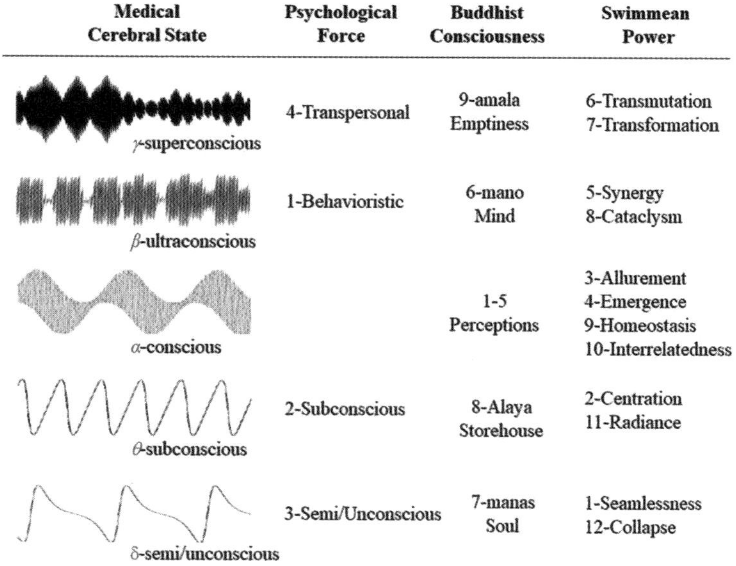

Medical Cerebral State	**Psychological Force**	**Buddhist Consciousness**	**Swimmean Power**
γ-superconscious	4-Transpersonal	9-amala Emptiness	6-Transmutation 7-Transformation
β-ultraconscious	1-Behavioristic	6-mano Mind	5-Synergy 8-Cataclysm
α-conscious		1-5 Perceptions	3-Allurement 4-Emergence 9-Homeostasis 10-Interrelatedness
θ-subconscious	2-Subconscious	8-Alaya Storehouse	2-Centration 11-Radiance
δ-semi/unconscious	3-Semi/Unconscious	7-manas Soul	1-Seamlessness 12-Collapse

Figure 5.4. EEG waves integrated with psychology, mind, and cosmogenetic powers

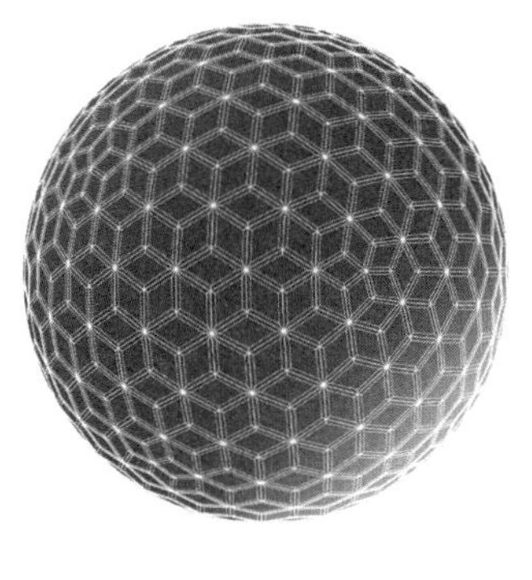

6 Holographic Theory of Everything—A Conclusion

This book presents in detail the holographic unity of heaven, earth, and humanity. It synthesizes Eastern wisdom, Western philosophy, and modern cosmology. The features of the three-tier holography are in conformance with the neo-Confucian world-ordering system of numerological cosmology, Swimmean cosmogenetic evolutionism, the Einstein-Friedmann dynamics of cosmology, the Hawking-Penrose theory of quantum gravity, and the Penrose-Hameroff cosmic quantum consciousness. Figure 6.1 illustrates the structure of the holography.

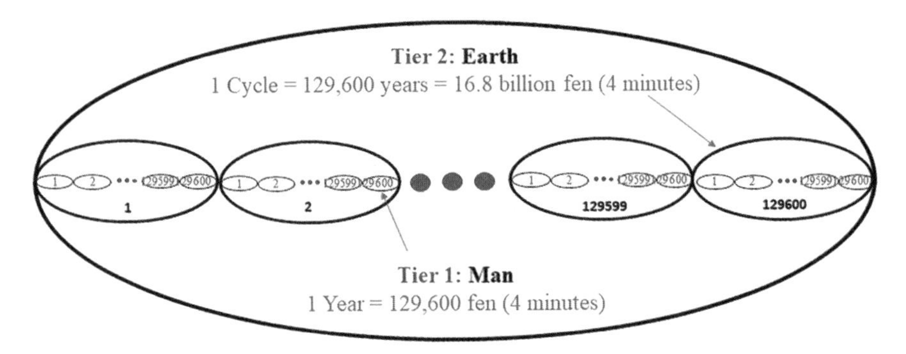

Figure 6.1. Structure of a three-tier holography

It describes that microcosmic man is indeed a holographic mode of the mesocosmic earth, and also that of the macrocosmic heaven. That is to say, humanity is inextricably and holistically blended in the cosmological evolution of nature, and able to unfold the mysteries of the universe at different tiers by way of mental activities. One such mystery lies in the holographic principles of the evolving entities at any tier. These principles provide quantitative answers to Einstein's quest for the theory of everything (ToE). They are described as follows.

Principle 1: The holography of heaven, earth, and humankind is represented by a horn-torus topology in a 3-D absolute space or an abridged 4-D curved space-time. This topology is perceived on a 2-D plane as the taiji diagram, where the small *yin* or *yang* dot in one half emerges due to

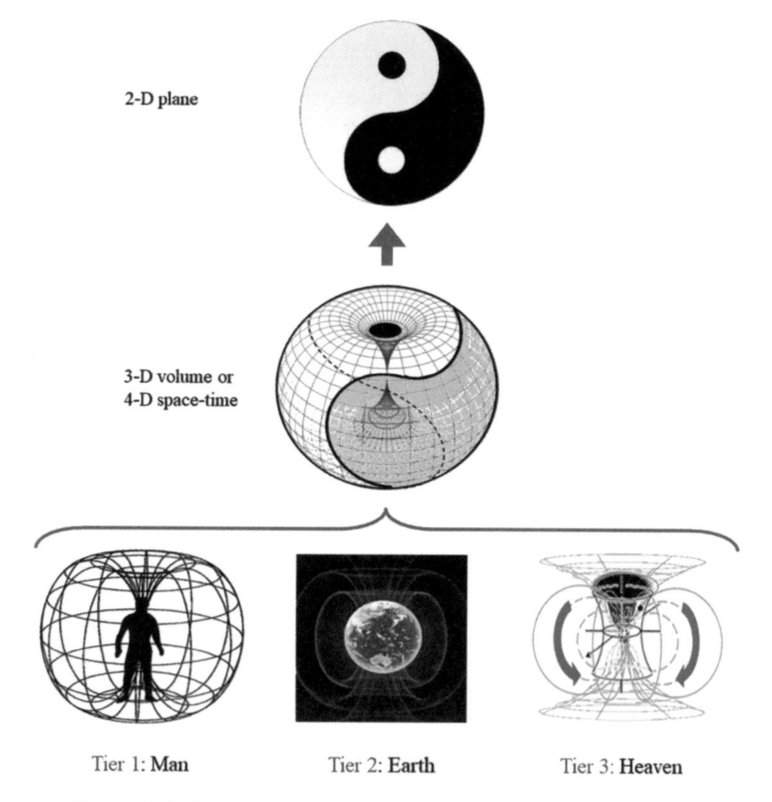

Figure 6.2. The three-tier holography as a 3-D horn torus and a 2-D taiji diagram

the curvature effects. The interdependence and mutual transformation of the *yin* and *yang* halves are concealed at at the center of the torus (fig. 6.2).

Principle 2: All tiers evolve in cycles as they follow Leibniz's binary (or Fuxi's) codes of the I Ching on a holographic torus topology (hereafter named I-torus). The topology has seven layers: latitudes at 30°, 60°, and 90° for both the upper and lower hemispheres, and equatorial 0°. Figure 6.3 shows the distributions of the I Ching's hexagrammatic *qi*, where the 12 sovereign stages are highlighted in red, forming dashed taiji lines. The distribution rule is provided in the appendix.

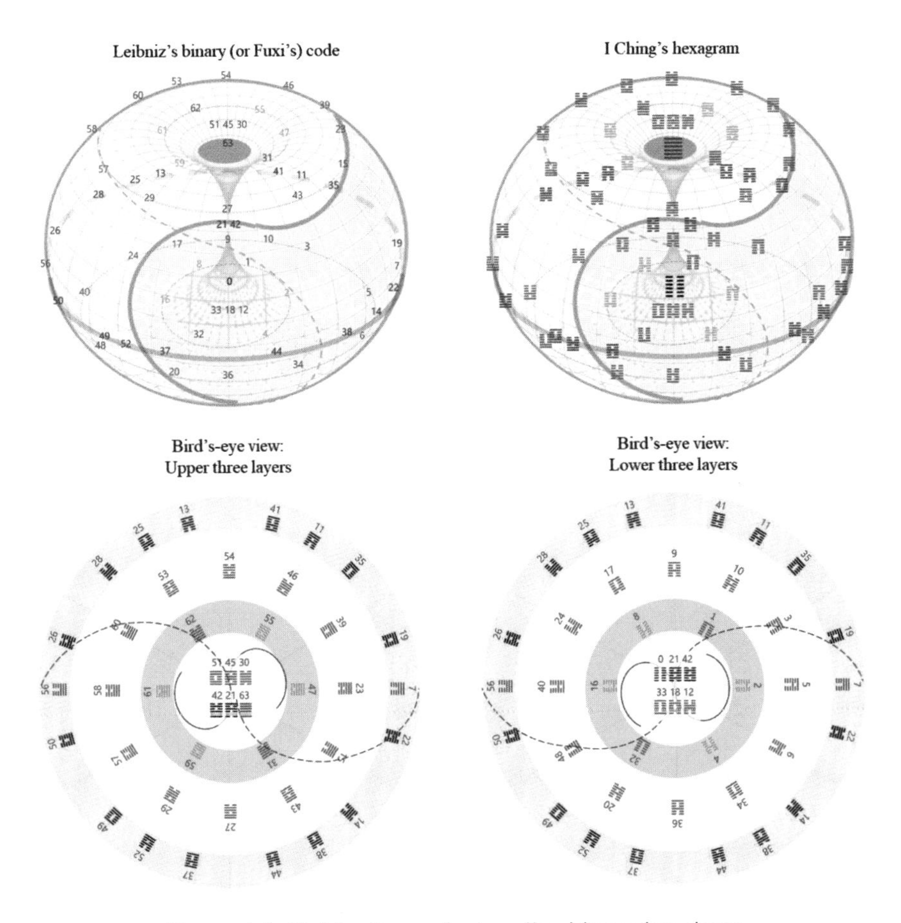

Figure 6.3. Distributions of *qi* on the I-torus topology

Principle 3: Every tier has its 129,600 units adhering to the I-torus with a scale-free sequential infrastructure holographically shared across all tiers, the five pillars of which alternate, in steps, between 12 and 30 numerical categories. From its inception, every unit has already possessed its evolutionary roadmap delineated by the related verses of the I Ching. Figure 6.4 illustrates the main steps accessing the mandate, using the year 2024 as an example.

Principle 4: Among the 12 epochal stages of every tier, around the midpoint of the third stage is when everything in the tier begins to individuate after coming into being. Meanwhile, the midpoint of the penultimate stage marks the time when everything starts to wither and return to its root. Consequently, approximately eight stages are left for the full evolution, as indicated by the distance between the two shaded texts in yellow in Figure 6.4.

Principle 5: The I-torus is characterized by a cosmic DNA, akin to the biological counterpart but operating across scale-free dimensions, governing the evolutionary path of natural entities. It is composed of 23 distinct codes, encompassing 12 Swimmean powers and 12 sovereign hexagrammatic *qi* (in alignment with the 12 terrestrial branches). At the center of the I-torus, two extreme *yin* and *yang* merge into each other to form the central point as a unit.

Hence, there are a total of 23 such codes. Figure 6.5 (p. 170) illustrates these codes as projected on the 2-D meridian plane of the I-torus. Table 6.1 (p. 171) lists these codes in the 3-D I-torus topology corresponding to the torus latitudes and longitudes, in addition to the solar terms and ecliptic angles in the solar ecliptic plane. Interestingly, within the human body, although there should be 24 pairs of chromosomes, each pair consists of two existing ones fusing end to end, resulting in 23 remaining.

In conclusion, we elucidate a holographic theory of everything (ToE) as a response to Einstein's enduring quest. It intricately weaves together

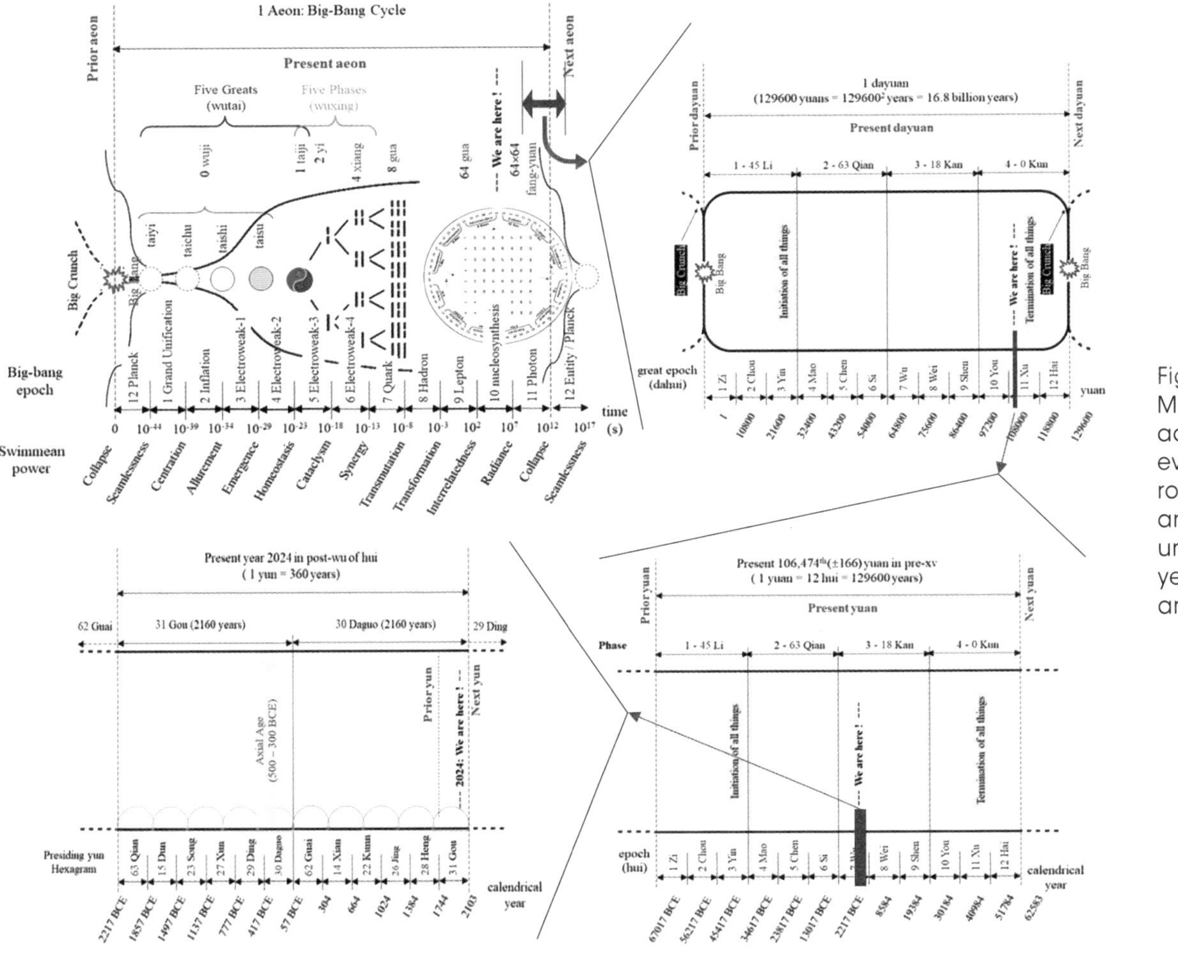

Figure 6.4. Main steps accessing the evolutionary roadmap of any arbitrary unit (using the year 2024 as an example)

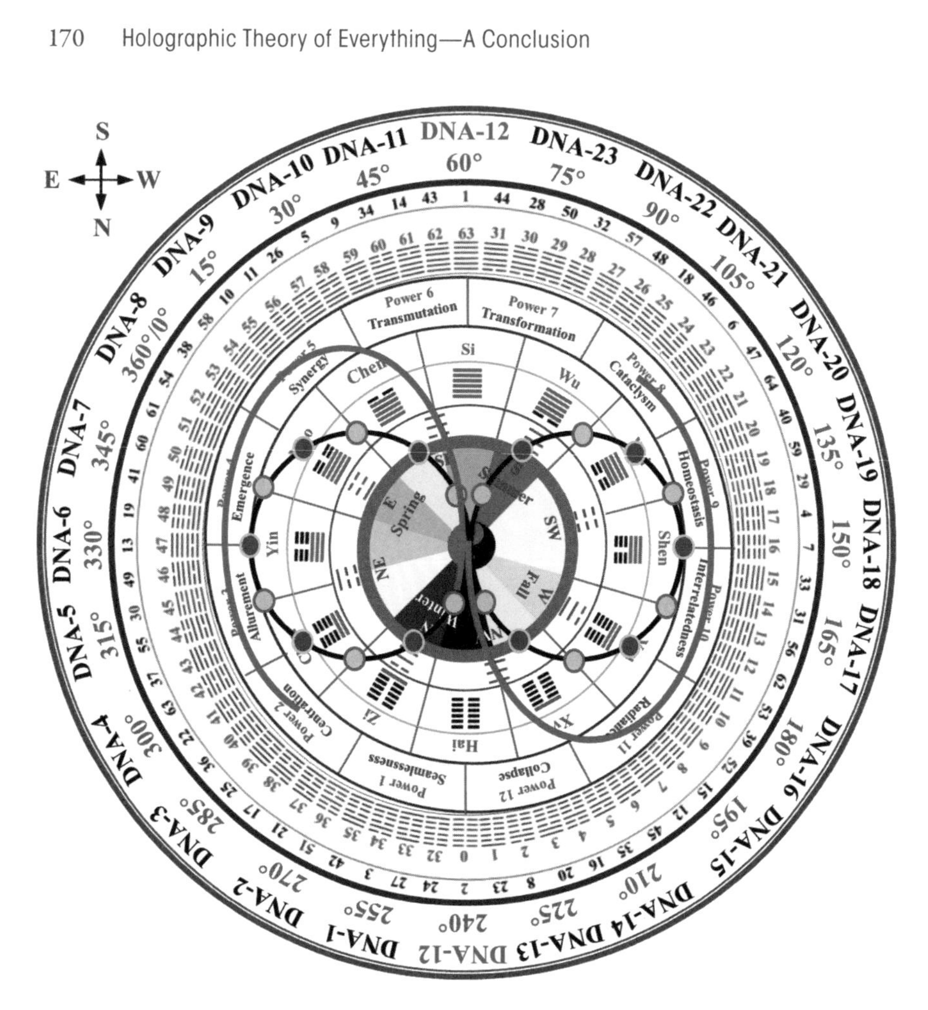

Figure 6.5. Holographic DNA codes projected on the 2-D meridian plane of the I-torus

the strands of Eastern wisdom, Western philosophy, and modern cosmology, unveiling the profound unity of heaven, earth, and humanity. Rooted in the tapestry of both Eastern and Western accomplishments, the theory reveals the interconnectedness of the macrocosmic heavens, mesocosmic Earth, and microcosmic humans.

As humanity actively engages in the cosmic ballet of evolution, our mental endeavors unfold the enigmatic mysteries of the universe across diverse tiers. At the heart of these revelations lie the five key principles, guiding our understanding of the cosmos. While reflecting

TABLE 6.1. HOLOGRAPHIC DNA CODES WITHIN THE 3-D I-TORUS TOPOLOGY

DNA		Swimmean Power	I Ching's *Qi*	Solar Term	Ecliptic		Torus			
							Latitude		Longitude	
12		Collapse		Winter Beginning	225°		75°N		15°W	
	12		䷁ Hai			240°		90°N		0°
				Less Snow						
1		Seamlessness			255°		75°N		15°E	
				Great Snow						
	2		䷗ Zi			270°		60°N		30°E
				Winter Solstice						
3		Centration			285°		45°N		45°E	
				Less Cold						
	4		䷒ Chou			300°		30°N		60°E
				Great Cold						
5		Allurement			315°		15°N		75°E	
				Spring Beginning						
	6		䷊ Yin			330°		0°		90°E
				Rain Water						
7		Emergence			345°		15°S		105°E	
				Insect Waking						
	8		䷡ Mao			360° (0°)		30°S		120°E
				Spring Equinox						
9		Synergy			15°		45°S		135°E	
				Pure Brightness						
	10		䷪ Chen			30°		60°S		150°E
				Grain Rain						
11		Transmutation			45°		75°S		165°E	
				Summer Beginning						
	12		䷀ Si			60°		90°S		180°E
				Less Fullness						
23		Transformation			75°		75°S		165°W	
				Grain in Ear						
	22		䷫ Wu			90°		60°S		150°W
				Summer Solstice						
21		Cataclysm			105°		45°S		135°W	
				Less Heat						
	20		䷠ Wei			120°		30°S		120°W
				Great Heat						
19		Homeostasis			135°		15°S		105°W	
				Autumn Beginning						
	18		䷋ Shen			150°		0°		90°W
				Heat End						
17		Interrelatedness			165°		15°N		75°W	
				White Dew						
	16		䷓ You			180°		30°N		60°W
				Autumn Equinox						
15		Radiance			195°		45°N		45°W	
				Cold Dew						
	14		䷖ Xv			210°		60°N		30°W
				Frost Descending						
13		Collapse			225°		75°N		15°W	
				Winter Beginning						
	12		䷁ Hai			240°		90°N		0°
12		Seamlessness		Less Snow	255°		75°N		15°E	

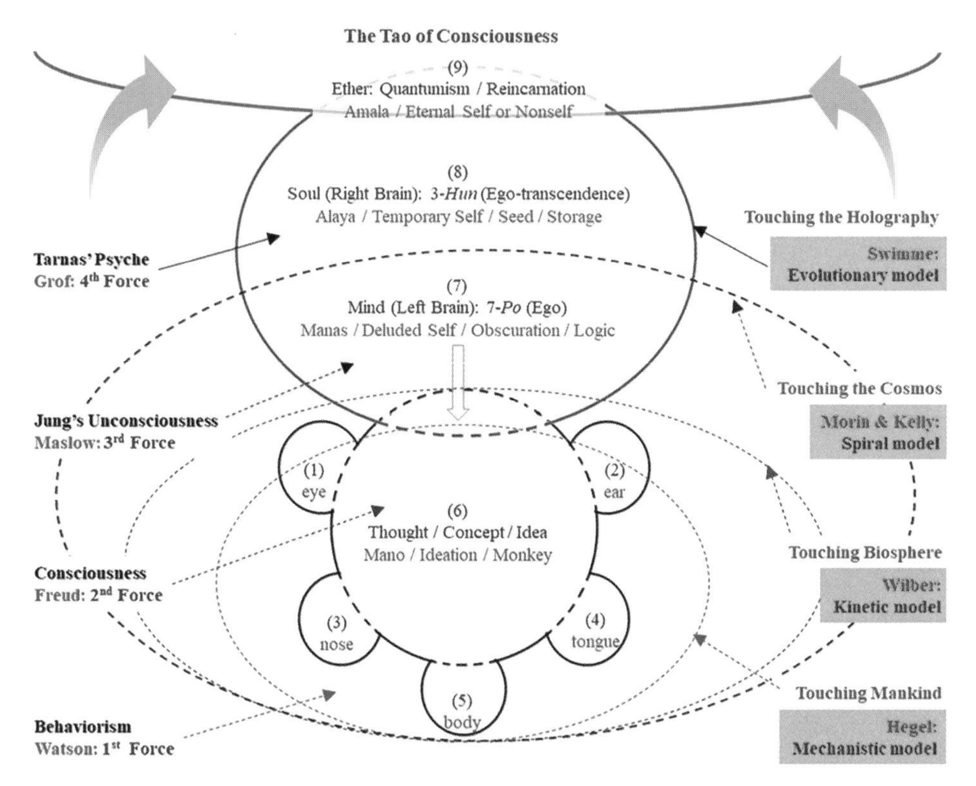

Figure 6.6. Overview of humans' transcendence in consciousness

on these insights, we embark on a transformative journey, transcending the limits of conventional understandings that define our existence. Figure 6.6 provides such an overview of humans' such continuous transcendence in consciousness toward the expansive dimensions of nature. Inviting contemplation of boundless mysteries, we need to go beyond the confines of the existing *uni*verse and consider our possible places in a *multi*verse. In our next book, we will delve into the quantum principles related to reincarnation, further expanding our exploration into the cosmic tapestry.

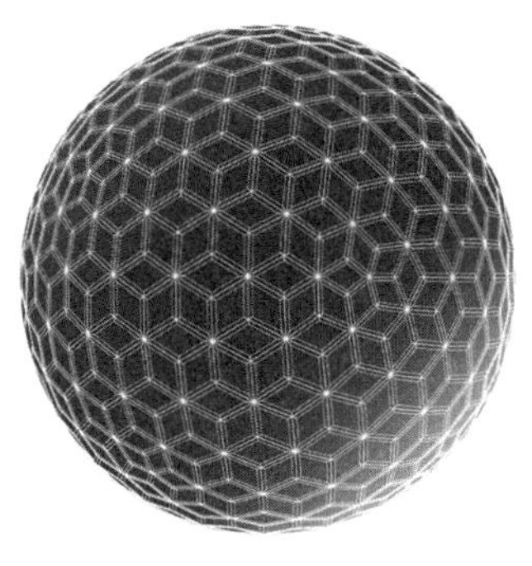

APPENDIX A

A Representative Set of 325 Cycles

This appendix is a selected set of 325 Cycles* that accommodates the Planck observational error of the age of our universe (13.799 ± 0.021 billion years). This set provides an outline for reading the full catalog of 129,600 Cycles as created by following the rules outlined in the table. For readers interested in the evolutionary development of any Cycle, find the number of the Cycle in the leftmost column and check the hexagram of the Cycle in the rightmost column. Afterward, read the related verses of the hexagram in the I Ching to understand the mandate of the Cycle.

S_5: Cycle No.	S_3: Great Revolution		S_4: Great Generation		Sexagenary Period	S_5: Cycle Hexagram
	Celestial Stem	Hexagram	Terrestrial Branch	Hexagram		
106312	Ji 296	57 Xun	Mao 3544	53 Jian	Yi Mao	64 Weiji
106313	Ji 296	57 Xun	Mao 3544	53 Jian	Bing Chen	40 Xie
106314	Ji 296	57 Xun	Mao 3544	53 Jian	Ding Si	59 Huan
106315	Ji 296	57 Xun	Mao 3544	53 Jian	Wu Wu	4 Meng
106316	Ji 296	57 Xun	Mao 3544	53 Jian	Ji Wei	7 Shi
106317	Ji 296	57 Xun	Mao 3544	53 Jian	Geng Shen	33 Dun
106318	Ji 296	57 Xun	Mao 3544	53 Jian	Xin You	31 Xian

*Energized by Power 11 in the Great Epoch (S_2) of the big-bang Great Cycle (S_1) of 129,600 Cycles.

S_5: Cycle No.	S_3: Great Revolution		S_4: Great Generation		Sexagenary Period	S_5: Cycle Hexagram
	Celestial Stem	Hexagram	Terrestrial Branch	Hexagram		
106320	Ji 296	57 Xun	Mao 3544	53 Jian	Gui Hai	62 Xiaoguo
106321	Ji 296	57 Xun	Chen 3545	59 Huan	Jia Zi	59 Huan
106322	Ji 296	57 Xun	Chen 3545	59 Huan	Yi Chou	4 Meng
106323	Ji 296	57 Xun	Chen 3545	59 Huan	Bing Yin	7 Shi
106324	Ji 296	57 Xun	Chen 3545	59 Huan	Ding Mao	33 Dun
106325	Ji 296	57 Xun	Chen 3545	59 Huan	Wu Chen	31 Xian
106326	Ji 296	57 Xun	Chen 3545	59 Huan	Ji Si	56 Lüü
106327	Ji 296	57 Xun	Chen 3545	59 Huan	Geng Wu	62 Xiaoguo
106328	Ji 296	57 Xun	Chen 3545	59 Huan	Xin Wei	53 Jian
106329	Ji 296	57 Xun	Chen 3545	59 Huan	Ren Shen	39 Jian
106330	Ji 296	57 Xun	Chen 3545	59 Huan	Gui You	52 Gen
106331	Ji 296	57 Xun	Chen 3545	59 Huan	Jia Xu	15 Qiann
106332	Ji 296	57 Xun	Chen 3545	59 Huan	Yi Hai	12 Pi
106333	Ji 296	57 Xun	Chen 3545	59 Huan	Bing Zi	45 Cui
106334	Ji 296	57 Xun	Chen 3545	59 Huan	Ding Chou	35 Jin
106335	Ji 296	57 Xun	Chen 3545	59 Huan	Wu Yin	16 Yv
106336	Ji 296	57 Xun	Chen 3545	59 Huan	Ji Mao	20 Guan
106337	Ji 296	57 Xun	Chen 3545	59 Huan	Geng Chen	8 Bi
106338	Ji 296	57 Xun	Chen 3545	59 Huan	Xin Si	23 Bo
106339	Ji 296	57 Xun	Chen 3545	59 Huan	Ren Wu	24 Fu
106340	Ji 296	57 Xun	Chen 3545	59 Huan	Gui Wei	27 Yi
106341	Ji 296	57 Xun	Chen 3545	59 Huan	Jia Shen	3 Zhun
106342	Ji 296	57 Xun	Chen 3545	59 Huan	Yi You	42 Yii
106343	Ji 296	57 Xun	Chen 3545	59 Huan	Bing Xu	51 Zhen
106344	Ji 296	57 Xun	Chen 3545	59 Huan	Ding Hai	21 Shihe
106345	Ji 296	57 Xun	Chen 3545	59 Huan	Wu Zi	17 Sui
106346	Ji 296	57 Xun	Chen 3545	59 Huan	Ji Chou	25 Wuwang
106347	Ji 296	57 Xun	Chen 3545	59 Huan	Geng Yin	36 Mingyi
106348	Ji 296	57 Xun	Chen 3545	59 Huan	Xin Mao	22 Bii
106349	Ji 296	57 Xun	Chen 3545	59 Huan	Ren Chen	63 Jiji
106350	Ji 296	57 Xun	Chen 3545	59 Huan	Gui Si	37 Jiaren
106351	Ji 296	57 Xun	Si 3546	59 Huan	Jia Wu	55 Feng
106352	Ji 296	57 Xun	Si 3546	59 Huan	Yi Wei	49 Ge

S_5: Cycle No.	S_3: Great Revolution		S_4: Great Generation		Sexagenary Period	S_5: Cycle Hexagram
	Celestial Stem	Hexagram	Terrestrial Branch	Hexagram		
106354	Ji 296	57 Xun	Si 3546	59 Huan	Ding You	19 Lin
106355	Ji 296	57 Xun	Si 3546	59 Huan	Wu Xu	41 Sun
106356	Ji 296	57 Xun	Si 3546	59 Huan	Ji Hai	60 Jie
106357	Ji 296	57 Xun	Si 3546	59 Huan	Geng Zi	61 Zhongfu
106358	Ji 296	57 Xun	Si 3546	59 Huan	Xin Chou	54 Guimei
106359	Ji 296	57 Xun	Si 3546	59 Huan	Ren Yin	38 Kui
106360	Ji 296	57 Xun	Si 3546	59 Huan	Gui Mao	58 Dui
106361	Ji 296	57 Xun	Si 3546	59 Huan	Jia Chen	10 Lü
106362	Ji 296	57 Xun	Si 3546	59 Huan	Yi Si	11 Tai
106363	Ji 296	57 Xun	Si 3546	59 Huan	Bing Wu	26 Daxu
106364	Ji 296	57 Xun	Si 3546	59 Huan	Ding Wei	5 Xu
106365	Ji 296	57 Xun	Si 3546	59 Huan	Wu Shen	9 Xiaoxu
106366	Ji 296	57 Xun	Si 3546	59 Huan	Ji You	34 Dazhuang
106367	Ji 296	57 Xun	Si 3546	59 Huan	Geng Xu	14 Dayou
106368	Ji 296	57 Xun	Si 3546	59 Huan	Xin Hai	43 Guai
106369	Ji 296	57 Xun	Si 3546	59 Huan	Ren Zi	44 Gou
106370	Ji 296	57 Xun	Si 3546	59 Huan	Gui Chou	28 Daguo
106371	Ji 296	57 Xun	Si 3546	59 Huan	Jia Yin	50 Ding
106372	Ji 296	57 Xun	Si 3546	59 Huan	Yi Mao	32 Heng
106373	Ji 296	57 Xun	Si 3546	59 Huan	Bing Chen	57 Xun
106374	Ji 296	57 Xun	Si 3546	59 Huan	Ding Si	48 Jing
106375	Ji 296	57 Xun	Si 3546	59 Huan	Wu Wu	18 Gu
106376	Ji 296	57 Xun	Si 3546	59 Huan	Ji Wei	46 Sheng
106377	Ji 296	57 Xun	Si 3546	59 Huan	Geng Shen	6 Song
106378	Ji 296	57 Xun	Si 3546	59 Huan	Xin You	47 Kun
106379	Ji 296	57 Xun	Si 3546	59 Huan	Ren Xu	64 Weiji
106380	Ji 296	57 Xun	Si 3546	59 Huan	Gui Hai	40 Xie
106381	Ji 296	57 Xun	Wu 3547	44 Gou	Jia Zi	44 Gou
106382	Ji 296	57 Xun	Wu 3547	44 Gou	Yi Chou	28 Daguo
106383	Ji 296	57 Xun	Wu 3547	44 Gou	Bing Yin	50 Ding
106384	Ji 296	57 Xun	Wu 3547	44 Gou	Ding Mao	32 Heng
106385	Ji 296	57 Xun	Wu 3547	44 Gou	Wu Chen	57 Xun
106386	Ji 296	57 Xun	Wu 3547	44 Gou	Ji Si	48 Jing

S_5: Cycle No.	S_3: Great Revolution		S_4: Great Generation		Sexagenary Period	S_5: Cycle Hexagram
	Celestial Stem	Hexagram	Terrestrial Branch	Hexagram		
106388	Ji 296	57 Xun	Wu 3547	44 Gou	Xin Wei	46 Sheng
106389	Ji 296	57 Xun	Wu 3547	44 Gou	Ren Shen	6 Song
106390	Ji 296	57 Xun	Wu 3547	44 Gou	Gui You	47 Kun
106391	Ji 296	57 Xun	Wu 3547	44 Gou	Jia Xu	64 Weiji
106392	Ji 296	57 Xun	Wu 3547	44 Gou	Yi Hai	40 Xie
106393	Ji 296	57 Xun	Wu 3547	44 Gou	Bing Zi	59 Huan
106394	Ji 296	57 Xun	Wu 3547	44 Gou	Ding Chou	4 Meng
106395	Ji 296	57 Xun	Wu 3547	44 Gou	Wu Yin	7 Shi
106396	Ji 296	57 Xun	Wu 3547	44 Gou	Ji Mao	33 Dun
106397	Ji 296	57 Xun	Wu 3547	44 Gou	Geng Chen	31 Xian
106398	Ji 296	57 Xun	Wu 3547	44 Gou	Xin Si	56 Lüü
106399	Ji 296	57 Xun	Wu 3547	44 Gou	Ren Wu	62 Xiaoguo
106400	Ji 296	57 Xun	Wu 3547	44 Gou	Gui Wei	53 Jian
106401	Ji 296	57 Xun	Wu 3547	44 Gou	Jia Shen	39 Jian
106402	Ji 296	57 Xun	Wu 3547	44 Gou	Yi You	52 Gen
106403	Ji 296	57 Xun	Wu 3547	44 Gou	Bing Xu	15 Qiann
106404	Ji 296	57 Xun	Wu 3547	44 Gou	Ding Hai	12 Pi
106405	Ji 296	57 Xun	Wu 3547	44 Gou	Wu Zi	45 Cui
106406	Ji 296	57 Xun	Wu 3547	44 Gou	Ji Chou	35 Jin
106407	Ji 296	57 Xun	Wu 3547	44 Gou	Geng Yin	16 Yv
106408	Ji 296	57 Xun	Wu 3547	44 Gou	Xin Mao	20 Guan
106409	Ji 296	57 Xun	Wu 3547	44 Gou	Ren Chen	8 Bi
106410	Ji 296	57 Xun	Wu 3547	44 Gou	Gui Si	23 Bo
106411	Ji 296	57 Xun	Wei 3548	44 Gou	Jia Wu	24 Fu
106412	Ji 296	57 Xun	Wei 3548	44 Gou	Yi Wei	27 Yi
106413	Ji 296	57 Xun	Wei 3548	44 Gou	Bing Shen	3 Zhun
106414	Ji 296	57 Xun	Wei 3548	44 Gou	Ding You	42 Yii
106415	Ji 296	57 Xun	Wei 3548	44 Gou	Wu Xu	51 Zhen
106416	Ji 296	57 Xun	Wei 3548	44 Gou	Ji Hai	21 Shihe
106417	Ji 296	57 Xun	Wei 3548	44 Gou	Geng Zi	17 Sui
106418	Ji 296	57 Xun	Wei 3548	44 Gou	Xin Chou	25 Wuwang
106419	Ji 296	57 Xun	Wei 3548	44 Gou	Ren Yin	36 Mingyi
106420	Ji 296	57 Xun	Wei 3548	44 Gou	Gui Mao	22 Bii

S_5: Cycle No.	S_3: Great Revolution		S_4: Great Generation		Sexagenary Period	S_5: Cycle Hexagram
	Celestial Stem	Hexagram	Terrestrial Branch	Hexagram		
106422	Ji 296	57 Xun	Wei 3548	44 Gou	Yi Si	37 Jiaren
106423	Ji 296	57 Xun	Wei 3548	44 Gou	Bing Wu	55 Feng
106424	Ji 296	57 Xun	Wei 3548	44 Gou	Ding Wei	49 Ge
106425	Ji 296	57 Xun	Wei 3548	44 Gou	Wu Shen	13 Tongren
106426	Ji 296	57 Xun	Wei 3548	44 Gou	Ji You	19 Lin
106427	Ji 296	57 Xun	Wei 3548	44 Gou	Geng Xu	41 Sun
106428	Ji 296	57 Xun	Wei 3548	44 Gou	Xin Hai	60 Jie
106429	Ji 296	57 Xun	Wei 3548	44 Gou	Ren Zi	61 Zhongfu
106430	Ji 296	57 Xun	Wei 3548	44 Gou	Gui Chou	54 Guimei
106431	Ji 296	57 Xun	Wei 3548	44 Gou	Jia Yin	38 Kui
106432	Ji 296	57 Xun	Wei 3548	44 Gou	Yi Mao	58 Dui
106433	Ji 296	57 Xun	Wei 3548	44 Gou	Bing Chen	10 Lü
106434	Ji 296	57 Xun	Wei 3548	44 Gou	Ding Si	11 Tai
106435	Ji 296	57 Xun	Wei 3548	44 Gou	Wu Wu	26 Daxu
106436	Ji 296	57 Xun	Wei 3548	44 Gou	Ji Wei	5 Xu
106437	Ji 296	57 Xun	Wei 3548	44 Gou	Geng Shen	9 Xiaoxu
106438	Ji 296	57 Xun	Wei 3548	44 Gou	Xin You	34 Dazhuang
106439	Ji 296	57 Xun	Wei 3548	44 Gou	Ren Xu	14 Dayou
106440	Ji 296	57 Xun	Wei 3548	44 Gou	Gui Hai	43 Guai
106441	Ji 296	57 Xun	Shen 3549	18 Gu	Jia Zi	18 Gu
106442	Ji 296	57 Xun	Shen 3549	18 Gu	Yi Chou	46 Sheng
106443	Ji 296	57 Xun	Shen 3549	18 Gu	Bing Yin	6 Song
106444	Ji 296	57 Xun	Shen 3549	18 Gu	Ding Mao	47 Kun
106445	Ji 296	57 Xun	Shen 3549	18 Gu	Wu Chen	64 Weiji
106446	Ji 296	57 Xun	Shen 3549	18 Gu	Ji Si	40 Xie
106447	Ji 296	57 Xun	Shen 3549	18 Gu	Geng Wu	59 Huan
106448	Ji 296	57 Xun	Shen 3549	18 Gu	Xin Wei	4 Meng
106449	Ji 296	57 Xun	Shen 3549	18 Gu	Ren Shen	7 Shi
106450	Ji 296	57 Xun	Shen 3549	18 Gu	Gui You	33 Dun
106451	Ji 296	57 Xun	Shen 3549	18 Gu	Jia Xu	31 Xian
106452	Ji 296	57 Xun	Shen 3549	18 Gu	Yi Hai	56 Lüü
106453	Ji 296	57 Xun	Shen 3549	18 Gu	Bing Zi	62 Xiaoguo
106454	Ji 296	57 Xun	Shen 3549	18 Gu	Ding Chou	53 Jian

S_5: Cycle No.	S_3: Great Revolution		S_4: Great Generation		Sexagenary Period	S_5: Cycle Hexagram
	Celestial Stem	Hexagram	Terrestrial Branch	Hexagram		
106456	Ji 296	57 Xun	Shen 3549	18 Gu	Ji Mao	52 Gen
106457	Ji 296	57 Xun	Shen 3549	18 Gu	Geng Chen	15 Qiann
106458	Ji 296	57 Xun	Shen 3549	18 Gu	Xin Si	12 Pi
106459	Ji 296	57 Xun	Shen 3549	18 Gu	Ren Wu	45 Cui
106460	Ji 296	57 Xun	Shen 3549	18 Gu	Gui Wei	35 Jin
106461	Ji 296	57 Xun	Shen 3549	18 Gu	Jia Shen	16 Yv
106462	Ji 296	57 Xun	Shen 3549	18 Gu	Yi You	20 Guan
106463	Ji 296	57 Xun	Shen 3549	18 Gu	Bing Xu	8 Bi
106464	Ji 296	57 Xun	Shen 3549	18 Gu	Ding Hai	23 Bo
106465	Ji 296	57 Xun	Shen 3549	18 Gu	Wu Zi	24 Fu
106466	Ji 296	57 Xun	Shen 3549	18 Gu	Ji Chou	27 Yi
106467	Ji 296	57 Xun	Shen 3549	18 Gu	Geng Yin	3 Zhun
106468	Ji 296	57 Xun	Shen 3549	18 Gu	Xin Mao	42 Yii
106469	Ji 296	57 Xun	Shen 3549	18 Gu	Ren Chen	51 Zhen
106470	Ji 296	57 Xun	Shen 3549	18 Gu	Gui Si	21 Shihe
106471	Ji 296	57 Xun	You 3550	18 Gu	Jia Wu	17 Sui
106472	Ji 296	57 Xun	You 3550	18 Gu	Yi Wei	25 Wuwang
106473	Ji 296	57 Xun	You 3550	18 Gu	Bing Shen	36 Mingyi
106474	Ji 296	57 Xun	You 3550	18 Gu	Ding You	22 Bii
106475	Ji 296	57 Xun	You 3550	18 Gu	Wu Xu	63 Jiji
106476	Ji 296	57 Xun	You 3550	18 Gu	Ji Hai	37 Jiaren
106477	Ji 296	57 Xun	You 3550	18 Gu	Geng Zi	55 Feng
106478	Ji 296	57 Xun	You 3550	18 Gu	Xin Chou	49 Ge
106479	Ji 296	57 Xun	You 3550	18 Gu	Ren Yin	13 Tongren
106480	Ji 296	57 Xun	You 3550	18 Gu	Gui Mao	19 Lin
106481	Ji 296	57 Xun	You 3550	18 Gu	Jia Chen	41 Sun
106482	Ji 296	57 Xun	You 3550	18 Gu	Yi Si	60 Jie
106483	Ji 296	57 Xun	You 3550	18 Gu	Bing Wu	61 Zhongfu
106484	Ji 296	57 Xun	You 3550	18 Gu	Ding Wei	54 Guimei
106485	Ji 296	57 Xun	You 3550	18 Gu	Wu Shen	38 Kui
106486	Ji 296	57 Xun	You 3550	18 Gu	Ji You	58 Dui
106487	Ji 296	57 Xun	You 3550	18 Gu	Geng Xu	10 Lü
106488	Ji 296	57 Xun	You 3550	18 Gu	Xin Hai	11 Tai

S_5: Cycle No.	S_3: Great Revolution		S_4: Great Generation		Sexagenary Period	S_5: Cycle Hexagram
	Celestial Stem	Hexagram	Terrestrial Branch	Hexagram		
106490	Ji 296	57 Xun	You 3550	18 Gu	Gui Chou	5 Xu
106491	Ji 296	57 Xun	You 3550	18 Gu	Jia Yin	9 Xiaoxu
106492	Ji 296	57 Xun	You 3550	18 Gu	Yi Mao	34 Dazhuang
106493	Ji 296	57 Xun	You 3550	18 Gu	Bing Chen	14 Dayou
106494	Ji 296	57 Xun	You 3550	18 Gu	Ding Si	43 Guai
106495	Ji 296	57 Xun	You 3550	18 Gu	Wu Wu	44 Gou
106496	Ji 296	57 Xun	You 3550	18 Gu	Ji Wei	28 Daguo
106497	Ji 296	57 Xun	You 3550	18 Gu	Geng Shen	50 Ding
106498	Ji 296	57 Xun	You 3550	18 Gu	Xin You	32 Heng
106499	Ji 296	57 Xun	You 3550	18 Gu	Ren Xu	57 Xun
106500	Ji 296	57 Xun	You 3550	18 Gu	Gui Hai	48 Jing
106501	Ji 296	57 Xun	Xu 3551	48 Jing	Jia Zi	48 Jing
106502	Ji 296	57 Xun	Xu 3551	48 Jing	Yi Chou	18 Gu
106503	Ji 296	57 Xun	Xu 3551	48 Jing	Bing Yin	46 Sheng
106504	Ji 296	57 Xun	Xu 3551	48 Jing	Ding Mao	6 Song
106505	Ji 296	57 Xun	Xu 3551	48 Jing	Wu Chen	47 Kun
106506	Ji 296	57 Xun	Xu 3551	48 Jing	Ji Si	64 Weiji
106507	Ji 296	57 Xun	Xu 3551	48 Jing	Geng Wu	40 Xie
106508	Ji 296	57 Xun	Xu 3551	48 Jing	Xin Wei	59 Huan
106509	Ji 296	57 Xun	Xu 3551	48 Jing	Ren Shen	4 Meng
106510	Ji 296	57 Xun	Xu 3551	48 Jing	Gui You	7 Shi
106511	Ji 296	57 Xun	Xu 3551	48 Jing	Jia Xu	33 Dun
106512	Ji 296	57 Xun	Xu 3551	48 Jing	Yi Hai	31 Xian
106513	Ji 296	57 Xun	Xu 3551	48 Jing	Bing Zi	56 Lüü
106514	Ji 296	57 Xun	Xu 3551	48 Jing	Ding Chou	62 Xiaoguo
106515	Ji 296	57 Xun	Xu 3551	48 Jing	Wu Yin	53 Jian
106516	Ji 296	57 Xun	Xu 3551	48 Jing	Ji Mao	39 Jian
106517	Ji 296	57 Xun	Xu 3551	48 Jing	Geng Chen	52 Gen
106518	Ji 296	57 Xun	Xu 3551	48 Jing	Xin Si	15 Qiann
106519	Ji 296	57 Xun	Xu 3551	48 Jing	Ren Wu	12 Pi
106520	Ji 296	57 Xun	Xu 3551	48 Jing	Gui Wei	45 Cui
106521	Ji 296	57 Xun	Xu 3551	48 Jing	Jia Shen	35 Jin
106522	Ji 296	57 Xun	Xu 3551	48 Jing	Yi You	16 Yv

S_5: Cycle No.	S_3: Great Revolution		S_4: Great Generation		Sexagenary Period	S_5: Cycle Hexagram
	Celestial Stem	Hexagram	Terrestrial Branch	Hexagram		
106524	Ji 296	57 Xun	Xu 3551	48 Jing	Ding Hai	8 Bi
106525	Ji 296	57 Xun	Xu 3551	48 Jing	Wu Zi	23 Bo
106526	Ji 296	57 Xun	Xu 3551	48 Jing	Ji Chou	24 Fu
106527	Ji 296	57 Xun	Xu 3551	48 Jing	Geng Yin	27 Yi
106528	Ji 296	57 Xun	Xu 3551	48 Jing	Xin Mao	3 Zhun
106529	Ji 296	57 Xun	Xu 3551	48 Jing	Ren Chen	42 Yii
106530	Ji 296	57 Xun	Xu 3551	48 Jing	Gui Si	51 Zhen
106531	Ji 296	57 Xun	Hai 3552	48 Jing	Jia Wu	21 Shihe
106532	Ji 296	57 Xun	Hai 3552	48 Jing	Yi Wei	17 Sui
106533	Ji 296	57 Xun	Hai 3552	48 Jing	Bing Shen	25 Wuwang
106534	Ji 296	57 Xun	Hai 3552	48 Jing	Ding You	36 Mingyi
106535	Ji 296	57 Xun	Hai 3552	48 Jing	Wu Xu	22 Bii
106536	Ji 296	57 Xun	Hai 3552	48 Jing	Ji Hai	63 Jiji
106537	Ji 296	57 Xun	Hai 3552	48 Jing	Geng Zi	37 Jiaren
106538	Ji 296	57 Xun	Hai 3552	48 Jing	Xin Chou	55 Feng
106539	Ji 296	57 Xun	Hai 3552	48 Jing	Ren Yin	49 Ge
106540	Ji 296	57 Xun	Hai 3552	48 Jing	Gui Mao	13 Tongren
106541	Ji 296	57 Xun	Hai 3552	48 Jing	Jia Chen	19 Lin
106542	Ji 296	57 Xun	Hai 3552	48 Jing	Yi Si	41 Sun
106543	Ji 296	57 Xun	Hai 3552	48 Jing	Bing Wu	60 Jie
106544	Ji 296	57 Xun	Hai 3552	48 Jing	Ding Wei	61 Zhongfu
106545	Ji 296	57 Xun	Hai 3552	48 Jing	Wu Shen	54 Guimei
106546	Ji 296	57 Xun	Hai 3552	48 Jing	Ji You	38 Kui
106547	Ji 296	57 Xun	Hai 3552	48 Jing	Geng Xu	58 Dui
106548	Ji 296	57 Xun	Hai 3552	48 Jing	Xin Hai	10 Lü
106549	Ji 296	57 Xun	Hai 3552	48 Jing	Ren Zi	11 Tai
106550	Ji 296	57 Xun	Hai 3552	48 Jing	Gui Chou	26 Daxu
106551	Ji 296	57 Xun	Hai 3552	48 Jing	Jia Yin	5 Xu
106552	Ji 296	57 Xun	Hai 3552	48 Jing	Yi Mao	9 Xiaoxu
106553	Ji 296	57 Xun	Hai 3552	48 Jing	Bing Chen	34 Dazhuang
106554	Ji 296	57 Xun	Hai 3552	48 Jing	Ding Si	14 Dayou
106555	Ji 296	57 Xun	Hai 3552	48 Jing	Wu Wu	43 Guai
106556	Ji 296	57 Xun	Hai 3552	48 Jing	Ji Wei	44 Gou

S_5: Cycle No.	S_3: Great Revolution		S_4: Great Generation		Sexagenary Period	S_5: Cycle Hexagram
	Celestial Stem	Hexagram	Terrestrial Branch	Hexagram		
106558	Ji 296	57 Xun	Hai 3552	48 Jing	Xin You	50 Ding
106559	Ji 296	57 Xun	Hai 3552	48 Jing	Ren Xu	32 Heng
106560	Ji 296	57 Xun	Hai 3552	48 Jing	Gui Hai	57 Xun
106561	Geng 297	20 Guan	Zi 3553	42 Yii	Jia Zi	42 Yii
106562	Geng 297	20 Guan	Zi 3553	42 Yii	Yi Chou	51 Zhen
106563	Geng 297	20 Guan	Zi 3553	42 Yii	Bing Yin	21 Shihe
106564	Geng 297	20 Guan	Zi 3553	42 Yii	Ding Mao	17 Sui
106565	Geng 297	20 Guan	Zi 3553	42 Yii	Wu Chen	25 Wuwang
106566	Geng 297	20 Guan	Zi 3553	42 Yii	Ji Si	36 Mingyi
106567	Geng 297	20 Guan	Zi 3553	42 Yii	Geng Wu	22 Bii
106568	Geng 297	20 Guan	Zi 3553	42 Yii	Xin Wei	63 Jiji
106569	Geng 297	20 Guan	Zi 3553	42 Yii	Ren Shen	37 Jiaren
106570	Geng 297	20 Guan	Zi 3553	42 Yii	Gui You	55 Feng
106571	Geng 297	20 Guan	Zi 3553	42 Yii	Jia Xu	49 Ge
106572	Geng 297	20 Guan	Zi 3553	42 Yii	Yi Hai	13 Tongren
106573	Geng 297	20 Guan	Zi 3553	42 Yii	Bing Zi	19 Lin
106574	Geng 297	20 Guan	Zi 3553	42 Yii	Ding Chou	41 Sun
106575	Geng 297	20 Guan	Zi 3553	42 Yii	Wu Yin	60 Jie
106576	Geng 297	20 Guan	Zi 3553	42 Yii	Ji Mao	61 Zhongfu
106577	Geng 297	20 Guan	Zi 3553	42 Yii	Geng Chen	54 Guimei
106578	Geng 297	20 Guan	Zi 3553	42 Yii	Xin Si	38 Kui
106579	Geng 297	20 Guan	Zi 3553	42 Yii	Ren Wu	58 Dui
106580	Geng 297	20 Guan	Zi 3553	42 Yii	Gui Wei	10 Lü
106581	Geng 297	20 Guan	Zi 3553	42 Yii	Jia Shen	11 Tai
106582	Geng 297	20 Guan	Zi 3553	42 Yii	Yi You	26 Daxu
106583	Geng 297	20 Guan	Zi 3553	42 Yii	Bing Xu	5 Xu
106584	Geng 297	20 Guan	Zi 3553	42 Yii	Ding Hai	9 Xiaoxu
106585	Geng 297	20 Guan	Zi 3553	42 Yii	Wu Zi	34 Dazhuang
106586	Geng 297	20 Guan	Zi 3553	42 Yii	Ji Chou	14 Dayou
106587	Geng 297	20 Guan	Zi 3553	42 Yii	Geng Yin	43 Guai
106588	Geng 297	20 Guan	Zi 3553	42 Yii	Xin Mao	44 Gou
106589	Geng 297	20 Guan	Zi 3553	42 Yii	Ren Chen	28 Daguo
106590	Geng 297	20 Guan	Zi 3553	42 Yii	Gui Si	50 Ding

S_5: Cycle No.	S_3: Great Revolution		S_4: Great Generation		Sexagenary Period	S_5: Cycle Hexagram
	Celestial Stem	Hexagram	Terrestrial Branch	Hexagram		
106592	Geng 297	20 Guan	Chou 3554	42 Yii	Yi Wei	57 Xun
106593	Geng 297	20 Guan	Chou 3554	42 Yii	Bing Shen	48 Jing
106594	Geng 297	20 Guan	Chou 3554	42 Yii	Ding You	18 Gu
106595	Geng 297	20 Guan	Chou 3554	42 Yii	Wu Xu	46 Sheng
106596	Geng 297	20 Guan	Chou 3554	42 Yii	Ji Hai	6 Song
106597	Geng 297	20 Guan	Chou 3554	42 Yii	Geng Zi	47 Kun
106598	Geng 297	20 Guan	Chou 3554	42 Yii	Xin Chou	64 Weiji
106599	Geng 297	20 Guan	Chou 3554	42 Yii	Ren Yin	40 Xie
106600	Geng 297	20 Guan	Chou 3554	42 Yii	Gui Mao	59 Huan
106601	Geng 297	20 Guan	Chou 3554	42 Yii	Jia Chen	4 Meng
106602	Geng 297	20 Guan	Chou 3554	42 Yii	Yi Si	7 Shi
106603	Geng 297	20 Guan	Chou 3554	42 Yii	Bing Wu	33 Dun
106604	Geng 297	20 Guan	Chou 3554	42 Yii	Ding Wei	31 Xian
106605	Geng 297	20 Guan	Chou 3554	42 Yii	Wu Shen	56 Lüü
106606	Geng 297	20 Guan	Chou 3554	42 Yii	Ji You	62 Xiaoguo
106607	Geng 297	20 Guan	Chou 3554	42 Yii	Geng Xu	53 Jian
106608	Geng 297	20 Guan	Chou 3554	42 Yii	Xin Hai	39 Jian
106609	Geng 297	20 Guan	Chou 3554	42 Yii	Ren Zi	52 Gen
106610	Geng 297	20 Guan	Chou 3554	42 Yii	Gui Chou	15 Qiann
106611	Geng 297	20 Guan	Chou 3554	42 Yii	Jia Yin	12 Pi
106612	Geng 297	20 Guan	Chou 3554	42 Yii	Yi Mao	45 Cui
106613	Geng 297	20 Guan	Chou 3554	42 Yii	Bing Chen	35 Jin
106614	Geng 297	20 Guan	Chou 3554	42 Yii	Ding Si	16 Yv
106615	Geng 297	20 Guan	Chou 3554	42 Yii	Wu Wu	20 Guan
106616	Geng 297	20 Guan	Chou 3554	42 Yii	Ji Wei	8 Bi
106617	Geng 297	20 Guan	Chou 3554	42 Yii	Geng Shen	23 Bo
106618	Geng 297	20 Guan	Chou 3554	42 Yii	Xin You	24 Fu
106619	Geng 297	20 Guan	Chou 3554	42 Yii	Ren Xu	27 Yi
106620	Geng 297	20 Guan	Chou 3554	42 Yii	Gui Hai	3 Zhun
106621	Geng 297	20 Guan	Yin 3555	59 Huan	Jia Zi	59 Huan
106622	Geng 297	20 Guan	Yin 3555	59 Huan	Yi Chou	4 Meng
106623	Geng 297	20 Guan	Yin 3555	59 Huan	Bing Yin	7 Shi
106624	Geng 297	20 Guan	Yin 3555	59 Huan	Ding Mao	33 Dun

S_5: Cycle No.	S_3: Great Revolution		S_4: Great Generation		Sexagenary Period	S_5: Cycle Hexagram
	Celestial Stem	Hexagram	Terrestrial Branch	Hexagram		
106626	Geng 297	20 Guan	Yin 3555	59 Huan	Ji Si	56 Lüü
106627	Geng 297	20 Guan	Yin 3555	59 Huan	Geng Wu	62 Xiaoguo
106628	Geng 297	20 Guan	Yin 3555	59 Huan	Xin Wei	53 Jian
106629	Geng 297	20 Guan	Yin 3555	59 Huan	Ren Shen	39 Jian
106630	Geng 297	20 Guan	Yin 3555	59 Huan	Gui You	52 Gen
106631	Geng 297	20 Guan	Yin 3555	59 Huan	Jia Xu	15 Qiann
106632	Geng 297	20 Guan	Yin 3555	59 Huan	Yi Hai	12 Pi
106633	Geng 297	20 Guan	Yin 3555	59 Huan	Bing Zi	45 Cui
106634	Geng 297	20 Guan	Yin 3555	59 Huan	Ding Chou	35 Jin
106635	Geng 297	20 Guan	Yin 3555	59 Huan	Wu Yin	16 Yv
106636	Geng 297	20 Guan	Yin 3555	59 Huan	Ji Mao	20 Guan

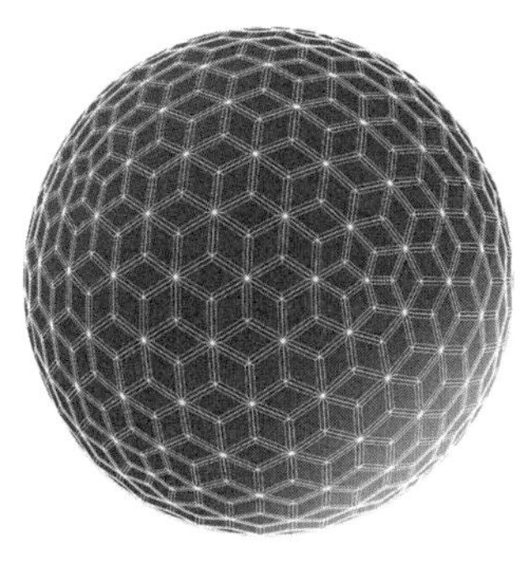

APPENDIX B

Rule of the Distributions of *Qi* on the I-torus Topology

This appendix outlines the constraints of the I Ching's hexagrammatic *qi* for distribution across the seven layers of the I-torus topology as shown in figure 6.3 (page 167). Considering the angle changes in latitude (L) from the lower hemisphere (H) to the upper one—from 90° lower H to 90° upper H—these layers contain the following array of hexagrams while most hexagrams are interdependent, relying on the summation of others in the distribution:

90°L lower H: 1 hexagram of 6 *yin* lines
60°L lower H: 6 hexagrams of 1 *yang* 5 *yin* lines
30°L lower H: 15 hexagrams of 2 *yang* 4 *yin* lines
0° equatorial: 20 hexagrams of 3 *yang* 3 *yin* lines
30°L upper H: 15 hexagrams of 4 *yang* 2 *yin* lines
60°L upper H: 6 hexagrams of 5 *yang* 1 *yin* lines
90°L upper H: 1 hexagram of 6 *yang* lines

Key of Table Elements

NB: Number of the neighbor hexagram single within the middle hexagrammatic circle

O: Number of the hexagram in the innermost circle opposite to the next closest hexagram within the same circle

D: Number of the hexagram on the same dashed line within the middle or the innermost circle

M1: Number of the vertically mirrored hexagram within the same circle

M2: Number of the *x*+45°-mirrored hexagram within the same circle

M3: Number of the *x*-45°-mirrored hexagram within the same circle

<table>
<tr><th rowspan="3">Fuxi's Code</th><th rowspan="3">Latitude (L)</th><th rowspan="3">Hemisphere (H)</th><th rowspan="3">Intermediary</th><th colspan="4">Summation</th></tr>
<tr><th colspan="3">Latitude (L)</th><th rowspan="2">Periodicity</th></tr>
<tr><th>30°</th><th colspan="2">60°</th></tr>
<tr><td>0</td><td colspan="2">Lower H and 0°L</td><td colspan="5" rowspan="3">-</td></tr>
<tr><td>1</td><td colspan="2">Lower H and 60°L</td></tr>
<tr><td>2</td><td colspan="2">Lower H and 60°L</td></tr>
<tr><td>3</td><td colspan="2">Lower H and 30°L</td><td colspan="2">-</td><td>1</td><td>2</td><td>-</td></tr>
<tr><td>4</td><td colspan="2">Lower H and 60°L</td><td colspan="5">-</td></tr>
<tr><td>5</td><td colspan="2">Lower H and 30°L</td><td>2 (D)</td><td>-</td><td>1</td><td>4</td><td rowspan="3">-</td></tr>
<tr><td>6</td><td colspan="2">Lower H and 30°L</td><td colspan="2">-</td><td>2</td><td>4</td></tr>
<tr><td rowspan="2">7</td><td rowspan="2">Equatorial 1</td><td>Lower</td><td rowspan="2">-</td><td>5 (D)</td><td colspan="2">2 (D)</td></tr>
<tr><td>Upper</td><td>23 (D)</td><td colspan="2">47 (D)</td><td>minus 63</td></tr>
<tr><td>8</td><td colspan="2">Lower H and 60°L</td><td colspan="5">-</td></tr>
<tr><td>9</td><td colspan="2">Lower H and 30°L</td><td colspan="2">-</td><td>1</td><td>8</td><td>-</td></tr>
<tr><td>10</td><td colspan="2">Lower H and 30°L</td><td>1 (D)</td><td>-</td><td>2</td><td>8</td><td></td></tr>
<tr><td rowspan="2">11</td><td rowspan="2">Equatorial 2</td><td>Lower</td><td>-</td><td>10 (D)</td><td colspan="2">1 (D)</td><td></td></tr>
<tr><td>Upper</td><td>38 (M1)</td><td>43 (D)</td><td colspan="2">31 (D)</td><td>minus 63</td></tr>
<tr><td>12</td><td colspan="2">Lower H and 30°L extra-1/3</td><td colspan="2">-</td><td>4</td><td>8</td><td rowspan="2">-</td></tr>
<tr><td rowspan="2">13</td><td rowspan="2">Equatorial 3</td><td>Lower</td><td>-</td><td>9 (NB)</td><td colspan="2">4 (O)</td></tr>
<tr><td>Upper</td><td>22 (M2)</td><td>15 (NB)</td><td colspan="2">61 (O)</td><td>minus 63</td></tr>
<tr><td rowspan="2">14</td><td rowspan="2">Equatorial 4</td><td>Lower</td><td rowspan="2">-</td><td>6 (NB)</td><td colspan="2">8 (O)</td><td>-</td></tr>
<tr><td>Upper</td><td>15 (NB)</td><td colspan="2">62 (O)</td><td rowspan="2">minus 63</td></tr>
<tr><td>15</td><td colspan="2">Upper H and 30°L</td><td colspan="2">-</td><td>31</td><td>47</td></tr>
</table>

<table>
<tr><th rowspan="3">Fuxi's Code</th><th rowspan="3">Latitude (L)</th><th rowspan="3">Hemisphere (H)</th><th rowspan="3">Intermediary</th><th colspan="4">Summation</th></tr>
<tr><th colspan="3">Latitude (L)</th><th rowspan="2">Periodicity</th></tr>
<tr><th>30°</th><th colspan="2">60°</th></tr>
<tr><td>16</td><td colspan="2">Lower H and 60°L</td><td colspan="5">-</td></tr>
<tr><td>17</td><td colspan="2">Lower H and 30°L</td><td colspan="2" rowspan="2">-</td><td>1</td><td>16</td><td rowspan="3">-</td></tr>
<tr><td>18</td><td colspan="2">Lower H and 30°L extra-2/3</td><td>2</td><td>16</td></tr>
<tr><td rowspan="2">19</td><td rowspan="2">Equatorial 5</td><td>Lower</td><td></td><td>3 (NB)</td><td colspan="2">16 (0)</td></tr>
<tr><td>Upper</td><td>37 (M3)</td><td>27 (NB)</td><td colspan="2">55 (0)</td><td>minus 63</td></tr>
<tr><td>20</td><td colspan="2">Lower H and 30°L</td><td colspan="2">-</td><td>4</td><td>16</td><td rowspan="2">-</td></tr>
<tr><td rowspan="2">21</td><td rowspan="2">Equatorial-1/2</td><td>Lower</td><td rowspan="3">-</td><td>5 / 17</td><td colspan="2">16 / 4</td></tr>
<tr><td>Upper</td><td>23 / 29</td><td colspan="2">61 / 55</td><td>minus 63</td></tr>
<tr><td rowspan="2">22</td><td rowspan="2">Equatorial 6</td><td>Lower</td><td>6 (NB)</td><td colspan="2">16 (0)</td><td>-</td></tr>
<tr><td>Upper</td><td>13 (M2)</td><td>54 (NB)</td><td colspan="2">31 (0)</td><td rowspan="2">minus 63</td></tr>
<tr><td>23</td><td colspan="2">Upper H and 30°L</td><td colspan="2" rowspan="2">-</td><td>31</td><td>55</td></tr>
<tr><td>24</td><td colspan="2">Lower H and 30°L</td><td>8</td><td>16</td><td rowspan="2">-</td></tr>
<tr><td rowspan="2">25</td><td rowspan="2">Equatorial 7</td><td>Lower</td><td>-</td><td>17 (D)</td><td colspan="2">8 (D)</td></tr>
<tr><td>Upper</td><td>52 (M1)</td><td>29 (D)</td><td colspan="2">59 (D)</td><td>minus 63</td></tr>
<tr><td rowspan="2">26</td><td rowspan="2">Equatorial 8</td><td>Lower</td><td>-</td><td>24 (NB)</td><td colspan="2">2 (0)</td><td>-</td></tr>
<tr><td>Upper</td><td>44 (M2)</td><td>27 (NB)</td><td colspan="2">62 (0)</td><td rowspan="2">minus 63</td></tr>
<tr><td>27</td><td colspan="2">Upper H and 30°L</td><td>-</td><td>31</td><td colspan="2">59</td></tr>
<tr><td rowspan="2">28</td><td rowspan="2">Equatorial 9</td><td>Lower</td><td rowspan="2">-</td><td>24 (NB)</td><td colspan="2">4 (0)</td><td>-</td></tr>
<tr><td>Upper</td><td>60 (NB)</td><td colspan="2">31 (0)</td><td rowspan="3">minus 63</td></tr>
<tr><td>29</td><td colspan="2">Upper H and 30°L</td><td rowspan="2">-</td><td>31</td><td colspan="2">61</td></tr>
<tr><td>30</td><td colspan="2">Upper H and 30°L extra-1/3</td><td>31</td><td colspan="2">62</td></tr>
<tr><td>31</td><td colspan="2">Upper H and 60°L</td><td colspan="5" rowspan="2">-</td></tr>
<tr><td>32</td><td colspan="2">Lower H and 60°L</td></tr>
<tr><td>33</td><td colspan="2">Lower H and 30°L extra-3/3</td><td colspan="2" rowspan="2">-</td><td>1</td><td>32</td><td rowspan="3">-</td></tr>
<tr><td>34</td><td colspan="2">Lower H and 30°L</td><td>2</td><td>32</td></tr>
<tr><td rowspan="2">35</td><td rowspan="2">Equatorial 10</td><td>Lower</td><td rowspan="2">-</td><td>3 (NB)</td><td colspan="2">32 (0)</td></tr>
<tr><td>Upper</td><td>39 (NB)</td><td colspan="2">59 (0)</td><td>minus 63</td></tr>
<tr><td>36</td><td colspan="2">Lower H and 30°L</td><td colspan="2">-</td><td>4</td><td>32</td><td rowspan="2">-</td></tr>
<tr><td rowspan="2">37</td><td rowspan="2">Equatorial 11</td><td>Lower</td><td>-</td><td>36 (NB)</td><td colspan="2">1 (0)</td></tr>
<tr><td>Upper</td><td>19 (M3)</td><td>39 (NB)</td><td colspan="2">61 (0)</td><td>minus 63</td></tr>
<tr><td rowspan="2">38</td><td rowspan="2">Equatorial 12</td><td>Lower</td><td>-</td><td>34 (D)</td><td colspan="2">4 (D)</td><td>-</td></tr>
<tr><td>Upper</td><td>11 (M1)</td><td>46 (D)</td><td colspan="2">55 (D)</td><td>minus 63</td></tr>
</table>

Fuxi's Code	Latitude (L)	Hemisphere (H)	Intermediary	Summation: Latitude (L) 30°	Summation: Latitude (L) 60°		Summation: Periodicity
39	Upper H and 30°L		-		47	55	minus 63
40	Lower H and 30°L				8	32	-
41	Equatorial 13	Lower	-	9 (NB)	32 (0)		
		Upper	50 (M3)	57 (NB)	47 (0)		minus 63
42	Equatorial-2/2	Lower	-	34 / 40	8 / 2		
		Upper		46 / 58	59 / 47		minus 63
43	Upper H and 30°L		-		47	59	
44	Equatorial 14	Lower	-	36 (NB)	8 (0)		-
		Upper	26 (M2)	60 (NB)	47 (0)		minus 63
45	Upper H and 30°L extra-2/3		-		47	61	
46	Upper H and 30°L				47	62	
47	Upper H and 60°L		-				
48	Lower H and 30°L		-		16	32	-
49	Equatorial 15	Lower	-	48 (NB)	1 (0)		
		Upper		57 (NB)	55 (0)		minus 63
50	Equatorial 16	Lower		48 (NB)	2 (0)		-
		Upper	41 (M3)	54 (NB)	59 (0)		minus 63
51	Upper H and 30°L extra-3/3		-		55	59	
52	Equatorial 17	Lower	-	20 (D)	32 (D)		-
		Upper	25 (M1)	53 (D)	62 (D)		minus 63
53	Upper H and 30°L		-		55	61	
54					55	62	
55	Upper H and 60°L		-				
56	Equatorial 18	Lower	-	40 (D)	16 (D)		
		Upper		58 (D)	61 (D)		minus 63
57	Upper H and 30°L		-		59	61	
58	Upper H and 30°L					62	
59	Upper H and 60°L		-				
60	Upper H and 30°L		-		61	62	minus 63
61	Upper H and 60°L		-				
62	Upper H and 60°L						
63	Upper H and 0°L						

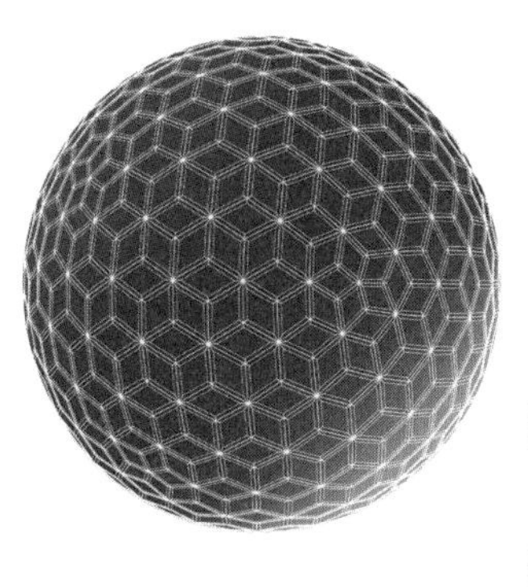

Compendium of Celebrities and Chinese Terms and Texts

TABLE C.1. CELEBRITIES OF THE EAST AND THE WEST

In alphabetical order of last names

(Note: Chinese names have their last name preceding the first name)

Name	Chinese Character	Pinyin	Life Span
Jean-Pierre Abel-Remusat	雷慕沙	léi mù shā	1788–1832
Aristarchus	阿里斯塔克斯	ā lǐ sī tǎ kè sī	310–230 BCE
Aristotle	亚里士多德	yà lǐ shì duō dé	384–322 BCE
Thomas Berry	托马斯•贝里	tuō mǎ sī · bèi lǐ	1914–2009
Brahma	梵天	fàn tiān	(Hindu) primordial
Cheng Hao (Chinese)	程颢	chéng hào	1032–1085
Cheng Yi (Chinese)	程颐	Chéng yí	1033–1107
Auguste Comte	奥古斯特•孔德	ào gǔ sī tè · kǒng dé	1798–1857
Copernicus	哥白尼	gē bái ní	1473–1543
Thomas Digges	托马斯•迪格斯	tuō mǎ sī · dí gé sī	1546–1595
Albert Einstein	阿尔伯特•爱因斯坦	ā ěr bó tè · aì yīn sī tǎn	1879–1955
Alexander Friedman	亚历山大•弗里德曼	yà lì shān dà · fú lǐ dé màn	1888–1925
Fuxi (Chinese)	伏羲	fú xī	ca. 2852–2737 BCE

TABLE C.1. CELEBRITIES OF THE EAST AND THE WEST (*cont.*)

Name	Chinese Character	Pinyin	Life Span
Gaia	盖亚	gāi yà	(Greek) primordial
Galileo Galilei	伽利略•伽利莱	jiā lì lüè · jiā lì lái	1564–1642
Stuart Hameroff	史都华•哈默洛夫	shǐ dōu huá · hā mò luò fū	1947–
James Hartle	詹姆斯•哈妥	zhān mǔ sī · hā tuǒ	1939–2023
Stephen Hawking	斯蒂芬•霍金	sī dì fēn · huò jīn	1942–2018
Georg Hegel	格奥尔格•黑格尔	gé ào ěr gé · hēi gé ěr	1770–1831
Carl Jung	卡尔•荣格	kǎ ěr · róng gé	1875–1961
Immanuel Kant	伊曼努尔•康德	yī màn nǔ ěr · kāng dé	1724–1804
Gottfried Leibniz	戈特弗里德•莱布尼茨	gě tè fú lǐ dé · lái bù ní zī	1646–1716
Heraclides	赫拉克里德	Hè lā kè lǐ dé	387–312 BCE
Hipparchus	喜帕恰斯	xǐ pà qià sī	190–120 BCE
Edwin Hubble	埃德温•哈勃	āi dé wēn · hā bó	1889–1953
Kant	康德	kāng dé	1724–1804
Johannes Kepler	约翰内斯•开普勒	yuē hàn nèi sī · kāi pǔ lè	1571–1630
Kongzi (or Confucius; Chinese)	孔子	Kǒng zǐ	551–479 BCE
Laozi (or Lao-tzu; Chinese)	老子	lǎo zǐ	ca. 571–471 BCE
Pierre-Simon Laplace	皮埃尔-西蒙•拉普拉斯	pí ài ěr - xī méng · lā pǔ lā sī	1749–1827
Georges Lemaître	乔治•勒梅特	qiáo zhì · lè méi tè	1894–1966
Liezi (Chinese)	列子	liè zǐ	450–375 BCE
James Maxwell	詹姆斯•麦克斯韦	zhān mǔ sī · mài kè sī wéi	1831–1879
John of Montecorvino	若望•孟高维诺	ruò wàng · mèng gāo wéi nuò	1247–1328
Robert Morrison	马礼逊	mǎ lǐ xùn	1782–1834
Nāgārjuna (Bodhisattva)	龙树(菩萨)	lóng shù (pú sà)	ca. 150–250
Isaac Newton	艾萨克•牛顿	ài sà kè · niú dùn	1642/3–1727
Nvwa (Chinese)	女娲	nǚ wā	(Chinese) primordial
Pangu (Chinese)	盘古	pán gǔ	(Chinese) primordial

TABLE C.1. CELEBRITIES OF THE EAST AND THE WEST (*cont.*)

Name	Chinese Character	Pinyin	Life Span
Plato	柏拉图	bó lā tú	ca. 427–347 BCE
Ptolemy	托勒密	tuō lè mì	100–170
Roger Penrose	罗杰•彭罗斯	luó jié · péng luó sī	1931–
Max Planck	马克斯•普朗克	mǎ kè sī · pǔ lǎng kè	1858–1947
Matteo Ricci	利玛窦	lì mǎ dòu	1552–1610
Timothy Richard	李提摩太	lǐ tí mó tài	1845–1919
Shakyamuni (Buddha)	释迦牟尼(佛)	Shì jiā móu ní (fó)	566–486 BCE
Shao Yong (Chinese)	邵雍	shào yōng	1011–1077
Willem de Sitter	威廉•德西特	wēi lián · dé xī tè	1872–1934
John Leighton Stuart	司徒雷登	sī tú léi dēng	1876–1962
Brian Swimme	布赖恩•斯威米	Bù lài ēn · sī wēi mǐ	1950–
Teilhard de Chardin	德日进	dé rì jìn	1881–1955
King Wen (of Zhou; Chinese)	(周)文王	(zhōu) wén wáng	1152–1056 BCE
Yao (Chinese)	尧	yáo	ca. 2377–2258 BCE
Zhang Zai (Chinese)	张载	zhāng zǎi	1020–1077
Zhou Dunyi (Chinese)	周敦颐	zhōu dūn yí	1017–1073
Zhu Xi (Chinese)	朱熹	zhū xī	1130–1200
Zhuangzi (Chinese)	庄子	zhuāng zǐ	ca. 369–286 BCE

TABLE C.2. CHINESE DYNASTIES

In chronological order

Term	Chinese Character	Pinyin	Reign Period
Xia dynasty	夏朝	xià cháo	2205–1766 BCE
Shang dynasty	商朝	shāng cháo	1523–1027 BCE
Zhou dynasty	周朝	zhōu cháo	1046–256 BCE
Warring States period	战国时期	zhàn guó shí qī	475–221 BCE

TABLE C.2. CHINESE DYNASTIES (*cont.*)

Term	Chinese Character	Pinyin	Reign Period
Qin dynasty	秦朝	qín cháo	221–206 BCE
Western Han dynasty	西汉	xī hàn	206 BCE–8
Tang dynasty	唐朝	táng cháo	618–907
Song dynasty	宋朝	sòng cháo	960–1279
Yuan dynasty	元朝	yuán cháo	1271–1368
Ming dynasty	明朝	míng cháo	1368–1644

TABLE C.3. CLASSICAL CHINESE WORKS

In alphabetical order

Term	Chinese Character	Pinyin	Translation
Dahuo lifa	大火历法	dà huǒ lì fǎ	Great Fire (α-Scorpii) Calendar
Huainanzi	淮南子	huái nán zǐ	Huainanzi
Huangdi yinfujing	黄帝阴符经	huáng dì yīn fú jīng	Yellow Emperor's Classic of the Secret Talisman
Huangji jingshi	皇极经世	huáng jí jīng shì	Supreme World-ordering Principles
Hunyuan bajing zhenjing	混元八景真经	hùn yuán bā jǐng zhēn jīng	Authentic Scripture on the Eight Effulgences of the Primordial Chaos
I Ching	易经	yì jīng	Book of Yi
Upper Ching	上经	shàng jīng	Upper Book
Lower Ching	下经	xià jīng	Lower Book
Liezi	列子	liè zǐ	Liezi
Limadou zhongwen zhuyiji	利玛窦中文著译集	lì mǎ dòu zhōng wén zhù yì jí	Matteo Ricci's Complete Corpus in Chinese
Shiyi	十翼	shí yì	Ten Wings (or Commentaries)
Taishang laojun shuo chang qingjing miaojing zuantu jiezhu	太上老君说常清静妙经纂图解注	tài shàng lǎo jūn shuō cháng qīng jìng miào jīng zuǎn tú jiě zhù	Explanations and Commentary with Diagrams to the Wondrous Canon of the Eternal Purity and Tranquility as Taught by the Supreme Venerable Sovereign
Tao Te Ching	道德经	dào dé jīng	Book of Dao and Virtue

TABLE C.3. CLASSICAL CHINESE WORKS (*cont.*)

Term	Chinese Character	Pinyin	Translation
Tianzhu Shiyi	天主实义	tiān zhǔ shí yì	The True Meaning of the Lord of Heaven
Wujing	五经	wǔ jīng	Five Classics
Xiali	夏历	xià lì	Xia Calendar
Yunji Qiqian	云笈七签	yún jí qī qiān	Seven Tablets in the Cloudy Satchel
Zhuangzi	庄子	zhuāng zǐ	Zhuangzi

TABLE C.4. CHINESE NUMEROLOGICAL SYMBOLISM

In numerical order

No.	Term	Chinese Character	Pinyin	Translation
0	wuji	无极	wú jí	*dao*; the ultimateless; ultimate reality
1	**taiji**	太极	**tài jí**	**oneness; the great ultimate**
	yin / yinqi	阴/阴气	yīn / yīn qì	receptive / feminine *qi*
	yang / yangqi	阳/阳气	yáng / yáng qì	creative / masculine *qi*
2	**liangyi**	两仪	**liǎng yí**	**two polarities**
	yinyi	阴仪	yīn yí	*yin* polarity
	yangyi	阳仪	yáng yí	*yang* polarity
2	**liangyao**	两爻	**liǎng yí**	**two polarities**
	yinyao	-- 阴爻	yīn yáo	broken line
	yangyao	— 阳爻	yáng yáo	solid line
3	**sanbao**	三宝	**sān bǎo**	**three bodily treasures (or jewels)**
	jing	精	jīng	body essence
	qi	气	qì	vital power
	shen	神	shén	conscious mind
4	**siqi**	(人体)四气	**(rén tǐ) sì qì**	**(bodily) four *qi***
	yuanqi	原气	yuán qì	inborn (or source) *qi*
	yingqi	营气	yíng qì	nutrient *qi*
	zongqi	宗气	zōng qì	pectoral *qi*
	weiqi	卫气	wèi qì	protective *qi*
4	**sixiang**	四象	**sì xiàng**	**four digrams (or symbols)**
	shaoyang	⚎ 少阳/木	shǎo yáng / mù	young *yang* / wood
	taiyang	⚌ 太阳/火	tài yáng / huǒ	mature *yang* / fire

TABLE C.4. CHINESE NUMEROLOGICAL SYMBOLISM (*cont.*)

No.	Term	Chinese Character	Pinyin	Translation
	shaoyin	⚍ 少阴/金	shǎo yīn / jīn	young *yin* / metal
	taiyin	⚏ 太阴/水	tài yīn / shuǐ	mature *yin* / water
4	**si zhenggua**	四正卦	**sì zhèng guà**	**four principal hexagrams**
	qian	䷀ 乾 1/63	qián	pure *yang* / heaven
	kun	䷁ 坤 2/0	kūn	pure *yin* / earth
	kan	䷜ 坎 29/18	kǎn	repeated sinking
	li	䷝ 离 30/45	lí	brightness
5	**(xiantian) wutai**	(先天)五太	**(xiān tiān) wǔ tài**	**five (primordial) greats**
	taiyi	太易	tài yì	great easiness (*qi* too weak to be seen)
	taichu	太初	tài chū	great origin - *qi* (*qi* emerging without shape xíng)
	taishi	太始	tài shǐ	great inaugural - *qi*, xíng (xíng emerging without quality zhì)
	taisu	太素	tài sù	great simplicity - *qi*, xíng, zhì (zhì emerging without entity tǐ)
	taiji	太极	tài jí	great ultimate - *qi*, xíng, zhì, tǐ (tǐ emerging to reach oneness)
5	**wuxing/wuqi** (sixiang + tianrun)	五行/五气 (四象+天润)	**wǔ xíng / wǔ qì** (sì xiàng + tiān rùn)	**five phases / five *qi*** (four digrams + heavenly nourishment)
	tianrun	☯ 天润/土	tiān rùn / tǔ	heavenly nourishment / earth
5	**wuxīng**	五星	**wǔ xīng**	**five (visible) planets**
	muxīng	木星	mù xīng	wood planet (Jupiter)
	huoxīng	火星	huǒ xīng	fire planet (Mars)
	jin-xīng	金星	jīn xīng"	metal planet (Venus)
	shuixīng	水星	shuǐ xīng	water planet (Mercury)
	tuxīng	土星	tǔ xīng	earth planet (Saturn)
5	**wuchang**	五常	**wǔ cháng**	**five (Confucian) virtues**
	ren	仁	rén	benevolence
	yi	义	yì	righteousness
	li	礼	lǐ	propriety
	zhi	智	zhì	wisdom
	xin	信	xìn	sincerity
5	**wuyi**	五易	**wǔ yì**	**five *yi***
	bu-yi	不易	bù yì	congruent *yi*

TABLE C.4. CHINESE NUMEROLOGICAL SYMBOLISM (*cont.*)

No.	Term	Chinese Character	Pinyin	Translation
	bian-yi	变易	biàn yì	forced self-*yi*
	jian-yi	简易	jiǎn yì	free self-*yi*
	he-yi	和易	hé yì	harmonic *yi*
	jiao-yi	交易	jiāo yì	interactive *yi*
6	**liuyao**	(卦的)六爻	**(guà de) liù yáo**	**(hexagrammatic) six lines**
	chuyao	初爻	chū yáo	bottom line
	eryao	二爻	èr yáo	seond line from the bottom
	sanyao	三爻	sān yáo	third line from the bottom
	siyao	四爻	sì yáo	fourth line from the bottom
	wuyao	五爻	wǔ yáo	fifth line from the bottom
	shangyao	上爻	shàng yáo	top line
6	**liuji**	(生命)六阶段	**(sheng mìng) liù jiē duàn**	**six stages (of life)**
	tai	胎	tāi	fetus
	sheng	生	shēng	production
	zhuang	壮	zhuàng	growth
	lao	老	lǎo	decay
	si	死	sǐ	destruction
	hua	化	huà	transformation
6	**(wuzhi qiyuan) liubu**	(物质起源)六步	**(wù zhì qǐ yuán) liù bù**	**six steps (of matter origin)**
	xìng	(物质属)性	(wù zhì shǔ) xìng	(matter's) property
	xíng	(物质)形(状)	(wù zhì) xíng (zhuàng)	(matter's) shape
	zhì	(物质品)质	(wù zhì pǐn) zhì	(matter's) quality
	tǐ	(物质实)体	(wù zhì shí) tǐ	(matter's) entity
	wù	(物质)本质	(wù zhì) běn zhì	(matter's) substance
	wù xìng	(物质)物性	(wù zhì) wù xìng	(matter's) attributes
8	**bagua**	八卦	**bā guà**	**eight trigrams**
	T1-qian	☰ 乾-天	qián - tiān	heaven
	T2-dui	☱ 兑-泽	duì - zé	lake
	T3-li	☲ 离-火	lí - huǒ	fire
	T4-zhen	☳ 震-雷	zhèn - léi	thunder
	T5-xun	☴ 巽-风	xùn - fēng	wind
	T6-kan	☵ 坎-水	kǎn - shuǐ	water
	T7-gen	☶ 艮-山	gèn - shān	mountain

TABLE C.4. CHINESE NUMEROLOGICAL SYMBOLISM (*cont.*)

No.	Term	Chinese Character	Pinyin	Translation
	T8-kun	☷ 坤–地	kūn - dì	earth
8	**ba jinggua**	八经卦	**bā jīng guà**	**eight superimposed hexagrams**
	qian	䷀ 乾 1/63	qián	pure *yang* / heaven
	dui	䷹ 兑 58/54	duì	joy
	li	䷝ 离 30/45	lí	brightness
	zhen	䷲ 震 51/36	zhèn	shaking
	xun	䷸ 巽 57/27	xùn	gentleness
	kan	䷜ 坎 29/18	kǎn	repeated sinking
	gen	䷳ 艮 52/9	gèn	stopping
	kun	䷁ 坤 2/0	kūn	pure *yin* / earth
10	**(wuxing zhi) shiqi**	(五行之)十气	**(wǔ xíng zhī) shí qì**	**(wuxing's) ten *qi***
	Q1-yangmu	阳木	yáng mù	*yang* wood
	Q2-yinmu	阴木	yīn mù	*yin* wood
	Q3-yanghuo	阳火	yáng huǒ	*yang* fire
	Q4-yinhuo	阴火	yīn huǒ	*yin* fire
	Q5-yangtu	阳土	yáng tǔ	*yang* earth
	Q6-yintu	阴土	yīn tǔ	*yin* earth
	Q7-yangjin	阳金	yáng jīn	*yang* metal
	Q8-yinjin	阴金	yīn jīn	*yin* metal
	Q9-yangshui	阳水	yáng shuǐ	*yang* water
	Q10-yinshui	阴水	yīn shuǐ	*yin* water
10	**shi tiangan**	十天干	**shí tiān gān**	**ten celestial stems**
	S1-jia	甲–破甲萌芽	jiǎ - pò jiǎ méng yá	continuous *qi* origination from void
	S2-yi	乙–柔弱初出	yǐ - róu ruò chū chū	weak growth
	S3-bing	丙–生长势旺	bǐng - sheng zhǎng shì wàng	vigorous growth
	S4-ding	丁–生长壮实	ding - sheng zhǎng zhuàng shí	substantial growth
	S5-wu	戊–生长成熟	wù - sheng zhǎng chéng shú	luxuriant ripening growth
	S6-ji	己–萎缩起始	jǐ - wěi suō qǐ shǐ	onset of perishment
	S7-geng	庚–枯萎衰亡	gēng - kū wěi shuāi wáng	withering to decline and fall
	S8-xin	辛–万象更新	xīn - wàn xiàng gēng xīn	preparation for a new turn
	S9-ren	壬–妊娠养育	rén - rèn shēn yang yù	preparation to hatch *qi*
	S10-gui	癸–原气初成	guǐ - yuan qì chū chéng	initial accomplishment of *qi*

TABLE C.4. CHINESE NUMEROLOGICAL SYMBOLISM (*cont.*)

No.	Term	Chinese Character	Pinyin	Translation
12	**shier dizhi**	十二地支	**shí èr dì zhī**	**twelve terrestrial branches**
	B1-zi	子-开	zǐ - kāi	origination
	B2-chou	丑-闭	chǒu - bì	preterm
	B3-yin	寅-建	yín - jiàn	initiation
	B4-mao	卯-除	mǎo - chú	replenishment
	B5-chen	辰-满	chén - mǎn	replenishment toward accomplishment
	B6-si	巳-平	sì - píng	removal of any over-replenishment
	B7-wu	午-定	wǔ - dìng	stabilization
	B8-wei	未-执	wèi - zhí	controlling
	B9-shen	申-破	shēn - pò	precariousness
	B10-you	酉-危	yǒu - wēi	collapse
	B11-xv	戌-成	xū - chéng	extinction
	B12-hai	亥-收	hài - shōu	completion
12	**shier zhugua**	十二主卦	**shí èr zhǔ guà**	**twelve sovereign hexagrams**
	H1-fu	䷗ 复 24/32	fù	returning
	H2-lin	䷒ 临 19/48	lín	approach
	H3-tai	䷊ 泰 11/56	tài	peace
	H4-dazhuang	䷡ 大壮 34/60	dà zhuàng	great strength
	H5-guai	䷪ 夬 43/62	guài	quick resolution
	H6-qian	䷀ 乾 1/63	qián	pure *yang* / heaven
	H7-gou	䷫ 姤 44/31	gòu	meeting
	H8-dun	䷠ 遯 33/15	dùn	retreat
	H9-pi	䷋ 否 12/7	pǐ	adversity
	H10-guan	䷓ 观 20/3	guān	contemplation
	H11-bo	䷖ 剥 23/1	bō	splitting apart
	H12-kun	䷁ 坤 2/0	kūn	pure *yin* / earth
12	**shier neizang**	十二内脏	**shí èr nèi zàng**	**twelve internal organs**
	O1-dan	胆	dǎn	gall bladder (GB)
	O2-gan	肝	gān	liver (LV)
	O3-fei	肺	fèi	lung (LU)
	O4-dachang	大肠	dà cháng	large intestine (LI)
	O5-wei	胃	wèi	stomach (ST)
	O6-pi	脾	pí	spleen (SP)

TABLE C.4. CHINESE NUMEROLOGICAL SYMBOLISM (*cont.*)

No.	Term	Chinese Character	Pinyin	Translation
	07-xin	心	xīn	heart (HT)
	08-xiaochang	小肠	xiǎo cháng	small intestine (SI)
	09-pangguang	膀胱	pāng guāng	urinary bladder (UB)
	010-shen	肾	shèn	kidney (KI)
	011-xinbao	心包	xīn bāo	pericardium (PC)
	012-sanjiao	三焦	sānjiāo	three burners (TB)
		上焦-胸	shàng jiao - xiōng	upper burner - chest
		中焦-腹	zhōng jiāo - fù	middle burner - abdominal
		下焦-盆	xià jiao - pén	lower burner - pelvic
12	**shier yinyuan**	十二因缘	**shí èr yīn yuán**	**twelve interdependent originations**
	wu-ming	无明	wú míng	ignorance
	xing	行	xíng	fabrication
	shi	(生之)识	(sheng zhī) shí	rebirth consciousness
	ming-se	名色	míng sè	name-form
	liu-ru	六入	liù rù	six-entrance
	chu	触	chù	contact
	shou	受	shòu	sense
	ai	爱	ài	craving
	qv	取	qǔ	attachment
	you	(易之)有	(yì zhī) yǒu	becoming to beingness
	sheng	生	shēng	birth
	lao-si	老死	lǎo sǐ	aging-deceasing
24	**ershisi jieqi**	二十四节(-中)气	**èr shí sì jié(-zhōng) qì**	**twenty-four nodal(-medial) *qi* / solar terms**
	J1-lichun	立春	lì chūn	spring beginning (SpB)
	J2-yushui	雨水	yǔ shuǐ	rain water (RW)
	J3-jingzhe	惊蛰	jīng zhé	insect waking (IW)
	J4-chunfen	春分	chūn fēn	spring equinox (SE)
	J5-qingming	清明	qīng míng	pure brightness (PB)
	J6-guyu	谷雨	gǔ yǔ	grain rain (GR)
	J7-lixia	立夏	lì xià	summer beginning (SuB)
	J8-xiaoman	小满	xiǎo mǎn	less fullness (LF)
	J9-mangzhong	芒种	máng zhòng	grain in ear (GE)
	J10-xiazhi	夏至	xià zhì	summer solstice (SS)

TABLE C.4. CHINESE NUMEROLOGICAL SYMBOLISM (*cont.*)

No.	Term	Chinese Character	Pinyin	Translation
	J11-xiaoshu	小暑	xiǎo shǔ	less heat (LH)
	J12-dashu	大暑	dà shǔ	great heat (GH)
	J13-liqiu	立秋	lì qiū	autumn beginning (AB)
	J14-chushu	处暑	chǔ shǔ	heat end (HE)
	J15-bailu	白露	bái lù	white dew (WD)
	J16-qiufen	秋分	qiū fēn	autumn equinox (AE)
	J17-hanlu	寒露	hán lù	cold dew (CD)
	J18-shuangjiang	霜降	shuāng jiàng	frost descending (FD)
	J19-lidong	立冬	lì dōng	winter beginning (WB)
	J20-xiaoxue	小雪	xiǎo xuě	less snow (LS)
	J21-daxue	大雪	dà xuě	great snow (GS)
	J22-dongzhi	冬至	dōng zhì	winter solstice (WS)
	J23-xiaohan	小寒	xiǎo hán	less cold (LC)
	J24-dahan	大寒	dà hán	great cold (GC)
64	**liushisi gua**	六十四卦	**liù shí sì guà**	**sixty-four hexagrams**
	qian	䷀ 乾 1/63	qián	pure *yang* / heaven / creativity
	kun	䷁ 坤 2/0	kūn	pure *yin* / earth / receptivity
	zhun	䷂ 屯 3/34	zhūn	initial difficulty
	meng	䷃ 蒙 4/17	méng	ignorance
	xu	䷄ 需 5/58	xū	waiting
	song	䷅ 讼 6/23	sòng	litigation
	shi	䷆ 师 7/16	shī	army
	bi	䷇ 比 8/2	bì	assistance
	xiaoxu	䷈ 小畜 9/59	xiǎo xù	small accumulation
	lv	䷉ 履* 10/55	lǚ	treading
	tai	䷊ 泰 11/56	tài	peace
	pi	䷋ 否 12/7	pǐ	adversity
	tongren	䷌ 同人 13/47	tóng rén	fellowship
	dayou	䷍ 大有 14/61	dà yōu	great possession
	qian	䷎ 谦 15/8	qiān	humility
	yv	䷏ 豫 16/4	yǜ	enjoyment
	sui	䷐ 随* 17/38	suí	following
	gu	䷑ 蛊 18/25	gǔ	work on decay
	lin	䷒ 临* 19/48	lín	approach

TABLE C.4. CHINESE NUMEROLOGICAL SYMBOLISM (*cont.*)

No.	Term	Chinese Character	Pinyin	Translation
	guan	䷓ 观 20/3	guān	contemplation
	shihe	䷔ 噬嗑 21/37	shì hé	biting to close up
	bi	䷕ 贲 22/41	bì	decoration
	bo	䷖ 剥 23/1	bō	splitting apart
	fu	䷗ 复 24/32	fù	returning
	wuwang	䷘ 无妄 25/39	wú wàng	non-fault
	daxv	䷙ 大畜 26/57	dà xù	great accumulation
	yi	䷚ 颐 27/33	yí	nourishment
	daguo	䷛ 大过* 28/30	dà guò	great passing
	kan	䷜ 坎 29/18	kǎn	repeated sinking
	li	䷝ 离 30/45	lí	brightness
	xian	䷞ 咸* 31/14	xián	response
	heng	䷟ 恒 32/28	héng	duration
	dun	䷠ 遁 33/15	dùn	retreat
	dazhuang	䷡ 大壮 34/60	dà zhuàng	great strength
	jin	䷢ 晋 35/5	jìn	going upward
	mingyi	䷣ 明夷 36/40	míng yí	injured brightness
	jiaren	䷤ 家人 37/43	jiā rén	family members
	kui	䷥ 睽* 38/53	kuí	opposition
	jian	䷦ 蹇 39/10	jiǎn	obstruction
	xie	䷧ 解 40/20	xiè	deliverance
	sun	䷨ 损* 41/49	sǔn	decrease
	yi	䷩ 益 42/35	yì	increase
	guai	䷪ 夬* 43/62	guài	quick resolution
	gou	䷫ 姤 44/31	gòu	meeting
	cui	䷬ 萃* 45/6	cuì	gathering
	sheng	䷭ 升 46/24	shēng	ascending
	kun	䷮ 困* 47/22	kùn	besetment
	jing	䷯ 井 48/26	jǐng	the well
	ge	䷰ 革* 49/46	gé	revolution
	ding	䷱ 鼎 50/29	dǐng	cauldron
	zhen	䷲ 震 51/36	zhèn	shaking
	gen	䷳ 艮 52/9	gèn	stopping

*15 hexagrams that Laozi used for the Tao Te Ching

TABLE C.4. CHINESE NUMEROLOGICAL SYMBOLISM (*cont.*)

No.	Term	Chinese Character	Pinyin	Translation
	jian	䷴ 渐 53/11	jiàn	gradual development
	guimei	䷵ 归妹* 54/52	guī mèi	young sister to marry
	feng	䷶ 丰 55/44	fēng	abundance
	lv	䷷ 旅 56/13	lǚ	wandering
	xun	䷸ 巽 57/27	xùn	gentleness
	dui	䷹ 兑* 58/54	duì	joy
	huan	䷺ 涣 59/19	huàn	dispersion
	jie	䷻ 节* 60/50	jié	restriction
	zhongfu	䷼ 中孚* 61/51	zhōng fú	inner sincerity
	xiaoguo	䷽ 小过 62/12	xiǎo guò	small passing
	jiji	䷾ 既济 63/42	jì jì	completion
	weiji	䷿ 未济 64/21	wèi jì	non-completion
	*15 hexagrams that Laozi used for the Tao Te Ching			

TABLE C.5. CHINESE WORDS

In alphabetical order of term

Term	Chinese Character	Pinyin	Translation
dao (or Tao)	道	dào	way, path, wuji, ultimate reality, one *yin* and one *yang*
li	理	lǐ	principle
qi	气	qì	(vital) power or energy
yang (binary 1)	阳 (二进制1)	yáng (èr jìn zhì yī)	masculine, positive, creative, active, bright, logical, warm, dynamic, expansive, projective, energetic, temporal, etc.
yi (or I)	易	yì	becoming, change, evolution
yin (binary 0)	阴 (二进制0)	yīn (èr jìn zhì líng)	feminine, negative, receptive, passive, dark, intuitive, cool, still, grounded, reflective, restful, spatial, etc.

In alphabetical order of translation

Term	Chinese Character	Pinyin	Translation
you	有	yǒu	beingness, existence, fullness
fojiao	佛教	fó jiào	Buddhism

TABLE C.5. CHINESE WORDS (*cont.*)

Term	Chinese Character	Pinyin	Translation
zhongyuan	中原	zhōng yuán	central plain
hundun	混沌	hùn dùn	chaos
jidujiao	基督教	jī dū jiào	Christianity
zhongyi	中医	zhōng yī	Chinese medicine
rujiao	儒教	rú jiào	Confucianism
neihan	内涵	nèi hán	(ontological) connotation
ke	克	kè	conquest
fengshui	风水	fēng shuǐ	cosmo-ecology
feng shui	风 水	fēng shuǐ	wind water
waiyan	外延	wài yán	(cosmological) denotation
daojiao	道教	dào jiào	Daoism
bagua	八卦	bā guà	eight trigrams
xian-tian-ba-gua hou-tian-ba-gua	先天八卦 后天八卦	xiān tiān bā guà hòu tiān bā guà	Prenatal/noumenal/anterior-heaven eight trigrams Postnatal/phenomenal/posterior-heaven eight trigrams
lì	力	lì	force
taishang	太上	tài shàng	grand supreme *qi*
wu	侮	wǔ	insult
zhi-run	置闰	zhì rùn	intercalation in timekeeping
yuánqǐ	缘起	yuán qǐ	interdependent origination (*pratītyasamutpāda*)
daoxue	道学	dào xué	learning of *dao*
lixue	理学	lǐ xué	learning of *li*
yao	爻	yáo	(hexagrammatic) line
jingluo	经络	jīng luò	meridian-submeridian
wanwu	万物	wàn wù	natural things
xukuo	虚廓	xū kuò	nebulous void
jieqi	节(-中)气	jié(-zhōng) qì	nodal(-medial) *qi* / solar terms
beisong wuzi	北宋五子	běi sòng wǔ zǐ	Northern Song five masters
wu	无	wú	nothingness, voidness, emptiness

TABLE C.5. CHINESE WORDS (*cont.*)

Term	Chinese Character	Pinyin	Translation
tianren heyi	天人合一	tiān rén hé yī	oneness of heaven and humanity
benti yuzhou lun	本体宇宙论	běn tǐ yǔ zhòu lùn	onto-cosmology
benti mingli xue	本体命理学	běn tǐ mìng lǐ xué	onto-numerology
houtian bagua	后天八卦	hòu tiān bā guà	posterior-heaven eight trigrams
yuanqi	元气/炁	yuán qì	primordial *qi*
sheng	生	shēng	production
kongxing	空性	kōng xìng	pure emptiness (*sunyata*)
zhigua	值卦	zhí guà	ruling hexagram
huhuang	惚恍	hū huǎng	seemingly visible but invisible
yí xī wēi	夷 希 微	yí xī wēi	colorless: invisible to seeing soundless: inaudible to hearing shapeless: intangible to touching
shikong	时空(宇宙)	shí kōng (yǔzhòu)	space-time (or universe)
fangyuantu	方圆图	fāng yuán tú	square-circular diagram
wushang	无上	wú shàng	supreme *qi*
xuanlao	玄老	xuán lǎo	supremely mysterious *qi*
shiwu	事物	shì wù	things
wùxing shénxing	物性 神性	wù xìng shén xìng	attributes/thingness (of things) rationality/unfathomability (of things)
kaiwu	开物	kāi wù	(things) to start
biwu	闭物	bì wù	(things) to terminate
shijian	时间	shí jiān	time
dayuan dahui dayun dashi yuan = danian	大元 大会 大运 大世 元=大年	dà yuán dà huì dà yùn dà shì yuán = dà nián	Great Cycle (129,600 Cycles) Great Epoch (10,800 Cycles); 1/12 of Great Cycle Great Revolution (360 Cycles); 1/30 of Great Epoch Great Generation (30 Cycles); 1/12 of Great Revolution Cycle = Great Year (129,600 years)
yuan hui yun shi nian = dafen	元 会 运 世 年=大分	yuán huì yùn shì nián = dà fēn	Cycle (129,600 years) Epoch (10,800 years); 1/12 of Cycle Revolution (360 years); 1/30 of Epoch Generation (30 years); 1/12 of Revolution Year = Great Fen (129,600 Fen)

TABLE C.5. CHINESE WORDS (*cont.*)

Term	Chinese Character	Pinyin	Translation
nian yue hou ri shichen fen	年 月 侯 日 时辰 分	nián yuè hòu rì shí chén fēn	Year; 1/30 of shi (Generation) Month; 1/12 of Year five-day (pentad) Day (6 *ri* = 6.0875 days); 1/30 of Month two-hour; 1/12 of Day four-minute
xvwu	虚无	xū wú	void
jingqi	景气	jǐng qì	void (or effulgence) *qi*
xiaoxi	消息	xiāo xī	waning and waxing
xiao xi	消 息	xiāo xī	waning (*yin*-ascending or *yang*-descending) waxing (*yang*-ascending or *yin*-descending)

TABLE C.6. CLASSICAL CHINESE PHRASES

In order by page number

Page		
Page 23	Phrase	独立不改，周行而不殆
	Pinyin	dú lì bù gǎi, zhōu xíng ér bù dài
	Translation	Revolves eternally without exhaustion, relying on nothing, with the end of one cycle to be the beginning of the next one
Page 24	Phrase	三才相盗
	Pinyin	sān cái xiāng dào
	Translation	Triadic Interdependence of Three Treasures (盗dao – interdependence or mutually making use of, rather than its literal meaning of "stealing" or "robbing")
Page 36	Phrase	类同相召，气同相合，声比则应
	Pinyin	lèi tóng xiāng zhào, qì tóng xiāng hé, shēng bǐ zé yìng
	Translation	Things of the same kind summon each other, those with the same *qi* coincide together, and those with matching sounds resonate
Page 36	Phrase	(易)无思也，无为也，寂然不动
	Pinyin	(yì) wú sī yě, wú wéi yě, jì rán bù dòng
	Translation	(*yi* owns) thoughtlessness, effortlessness, quietness, and stationariness
Pages 36–37	Phrase	(易能被)感而遂通天下之故
	Pinyin	(yì néng bèi) gǎn ér suí tōng tiān xià zhī gù
	Translation	(*yi* is able to be) induced to an activity producing responses that penetrate forthwith to all the phenomena and events of the universe upon stimuli

TABLE C.6. CLASSICAL CHINESE PHRASES (*cont.*)

Page		
Page 37	Phrase	一阴一阳之(称)谓道
	Pinyin	yī yīn yī yáng zhī (chēng) wèi dào
	Translation	The appellation of the unity of one *yin* and one *yang* is called the *dao*
Page 38	Phrase	无，名天下之始；有， 名万物之母 . . . 天下万物生于有，有生于无
	Pinyin	wú, míng tiān xià zhī shǐ; yǒu, míng wàn wù zhī mǔ . . . tiān xià wàn wù shēng yú yǒu, yǒu shēng yú wú
	Translation	Nothingness is named the Origination of all under Heaven; Beingness is named the Mother of every entity of reality . . . All the entities under Heaven are begotten from the Beingness, and the Beingness is begotten from the Nothingness
Page 38	Phrase	夫昭昭生于冥冥，有伦生于无形，精神生于道，形本生于精，而万物以形相生
	Pinyin	fū zhāo zhāo shēng yú míng míng, yǒu lún shēng yú wú xíng, jīng shén shēng yú dào, xíng běn shēng yú jīng, ér wàn wù yǐ xíng xiāng shēng
	Translation	The luminous is born from the obscure, the multiform from the unembodied, the spiritual from dao, the bodily from the seminal essence, and all things from the bodily of one another
Pages 38–39	Phrase	有实而无乎处者，宇也；有长而无本剽者，宙也；. . . 万物出乎无有；有不能以有为有；必出乎无有；而无有一无有
	Pinyin	yǒu shí ér wú hū chù zhě, yǔ yě; yǒu zhǎng ér wú běn piāo zhě, zhòu yě; . . . wàn wù chū hū wú yǒu; yǒu bù néng yǐ yǒu wéi yǒu; bì chū hū wú yǒu; ér wú yǒu yī wú yǒu
	Translation	What "has a real existence but has nothing to do with position is the cosmological space"; what "has a continuance but has nothing to do with either beginning or end is the cosmological time"; . . . "all things come from non-existence, but the (first) existences are unable to bring themselves into existence except for the non-being which is just the same as the non-existing (in reality)
Page 39	Phrase	道始生虚廓，虚廓生宇宙，宇宙生炁，炁有涯垠，清阳者薄靡而为天，重浊者凝滞而为地
	Pinyin	dào shǐ shēng xū kuò, xū kuò shēng yǔ zhòu, yǔ zhòu shēng qì, qì yǒu yá yín, qīng yáng zhě bó mǐ ér wéi tiān, zhòng zhuó zhě níng zhì ér wéi dì
	Translation	The *dao* produces nebulous voids; the void produces space-time; the space-time itself gives rise to the primordial *qi* which is enclosed by a horizon; within the horizon, the pure-bright part spreads out to give rise to Heaven, and the heavy-turbid one congeals to give rise to Earth
Pages 39–40	Phrase	夫天地以前 . . . 只是虚无；虚无之中有景炁(带着淡黄色)；景炁极而生杳冥，杳冥极方有润湿；润显之极，始结成雾露，雾露之极，方变水 . . . 流水者，阴气；阴极始生阳气 . . . 阴阳相炼，其数满足，始结为混沌；混沌 . . . 方为一气，一气所萌，方为天地之母

TABLE C.6. CLASSICAL CHINESE PHRASES (*cont.*)

Pages 39–40	**Pinyin**	fū tiān dì yǐ qián . . . zhǐ shì xū wú; xū wú zhī zhōng yǒu jīng qì (dài zhe dàn huáng sè); jīng qì jí ér shēng yǎo míng, yǎo míng jí fāng yǒu rùn shī; rùn xiǎn zhī jí, shǐ jié chéng wù lù, wù lù zhī jí, fāng biàn shuǐ . . . liú shuǐ zhě, yīn qì; yīn jí shǐ shēng yáng qì . . . yīn yáng xiāng liàn, qí shù mǎn zú, shǐ jié wéi hùn dùn; hùn dùn . . . fāng wéi yī qì, yī qì suǒ méng, fāng wéi tiān dì zhī mǔ
	Translation	Before heaven and earth . . . there is merely the Void where only jingqi resides (appearing as a yellowish hue); jingqi is so tenuous as to become a dim and dusky state when fluctuating to reach its utmost; when the state fluctuates to an extrame, moistening is nourished; the moistness went to form a fog in its extreme; the fog produces water in its extreme; . . . the flow of water gives rise to yinqi; at the extreme of yinqi, yangqi is produced . . . the mutual interactions of yinqi and yangqi constitute hundun, the primordial chaos, after sufficient times of reciprocal influences; the chaos . . . is the mixed *qi* which springs natural things as the mother of Heaven and Earth
Page 40	**Phrase**	于幽原之中而生一炁焉。 化生之后九十九万亿九十九万岁，乃化生三气。各相去九十九万亿九十九万岁，共生无上也； 自无上生后九十九万亿九十九万岁，乃生中二气也，中三气也；中二气、中三气各相去九十九万亿九十九万岁，三合 . . . 共成玄老也； 自玄老生后九十九万亿九十九万岁，乃化生下三气也；下三气各相去九十九万亿九十九万岁，三合 . . . 共成太上也
	Pinyin	yú yōu yuán zhī zhōng ér shēng yī qì yān huà shēng zhī hòu jiǔ shí jiǔ wàn yì jiǔ shí jiǔ wàn suì, nǎi huà shēng sān qì. Gè xiāng qù jiǔ shí jiǔ wàn yì jiǔ shí jiǔ wàn suì, gòng shēng wú shàng yě zì wú shàng sheng hòu jiǔ shí jiǔ wàn yì jiǔ shí jiǔ wàn suì, nǎi shēng zhōng èr qì yě, zhōng sān qì yě; zhōng èr qì, zhōng sān qì gè xiāng qù jiǔ shí jiǔ wàn yì jiǔ shí jiǔ wàn suì, sān hé . . . gòng chéng xuán lǎo yě zì xuán lǎo sheng hòu jiǔ shí jiǔ wàn yì jiǔ shí jiǔ wàn suì, nǎi huà shēng xià sān qì yě; xià sān qì gè xiāng qù jiǔ shí jiǔ wàn yì jiǔ shí jiǔ wàn suì, sān hé . . . gòng chéng tài shàng yě
	Translation	At first, a primordial *qi* emerges from the void After 99,000,000,990,000 years, the produced primordial *qi* transforms three times into three *qi*, respectively, with intervals of the same number of the years in-between; The resulting *qi* of these three transformations together produces the supreme *qi* named wushang In two intervals with the same number of the years after the supreme *qi* is produced, the second middle *qi* and the third middle *qi* come into being successively; The supreme *qi*, the second middle *qi*, and the third one together . . . produce the supremely mysterious *qi* named xuanlao In three intervals with the same number of years after the supremely mysterious *qi* is produced, the lower three *qi* are transformed and together . . . produce the grand supreme *qi* named taishang

TABLE C.6. CLASSICAL CHINESE PHRASES (*cont.*)

Page		
Page 41	**Phrase**	(混沌中)有一阳初动于中，便生奇偶，分阴分阳，生育天地 清气上腾而生成天，浊气下降而结成地. 阴气出地而复上升于天，阳气从天而复下降于地. 阴阳往来，循环不已 . . . 日月运行，五气顺布，四时行焉，故能长养万物
	Pinyin	(hùn dùn zhōng) yǒu yī yáng chū dòng yú zhōng, biàn shēng qí ǒu, fēn yīn fēn yáng, shēng yù tiān dì qīng qì shàng téng ér shēng chéng tiān, zhuó qì xià jiàng ér jié chéng dì, yīn qì chū dì ér fù shàng shēng yú tiān, yáng qì cóng tiān ér fù xià jiàng yú dì. yīn yáng wǎng lái, xún huán bù yǐ . . . rì yuè yùn xíng, wǔ qì shùn bù, sì shí xíng yān, gù néng cháng yǎng wàn wù
	Translation	Initially, a seed of *yang* becomes active (in the chaos) to generate odd and even; *yin* and *yang* are thus divided to give birth to heaven and earth The heaven is formed by the light and pure *qi* which rises upward to spread, while the earth is formed by the dense and impure *qi* which solidifies downward to condense; yinqi can become active to flow out of the earth and rise up to the heaven, while yangqi can become still to flow out of the heaven and condense down to the earth. *Yin* and *yang* alternate and circulate endlessly . . . to drive the celestial motion of the Sun and the Moon, the harmonic distribution of the five phases' *qi*, and the proceeding course of the four seasons; therefore, all the things are always able to be brought to maturity
Page 42	**Phrase**	天下之物生于有，有生于无
	Pinyin	tiān xià zhī wù shēng yú yǒu, yǒu shēng yú wú
	Translation	All the things originate from *you* (the beingness); and that you is begotten of *wu* (the nothingness, voidness, or emptiness)
Page 42	**Phrase**	道生一
	Pinyin	dào shēng yī
	Translation	*dao* begets the oneness
Page 42	**Phrase**	无极而太极
	Pinyin	wú jí ér tài jí
	Translation	From wuji, taiji comes to be
Page 42	**Phrase**	太极一也
	Pinyin	tài jí yī yě
	Translation	Taiji is nothing else but the oneness
Page 43	**Phrase**	冯冯翼翼，洞洞灟灟
	Pinyin	féng féng yì yì, dòng dòng zhuó zhuó
	Translation	ascending and flying, diving and delving

TABLE C.6. CLASSICAL CHINESE PHRASES (*cont.*)

Page		
Page 43	**Phrase**	道始生虚廓
	Pinyin	dào shǐ shēng xū kuò
	Translation	*dao* begins with the appearance of the nebulous voids
Page 43	**Phrase**	太易者，未见(虚廓生之弱)气也；太初者，气之始也[未见(物质之)形]；太始者，形之始也[未见(物质之)质]；太素者，质之始也[未见(物质之)体]
	Pinyin	tài yì zhě, wèi jiàn (xū kuò shēng zhī ruò) qì yě; tài chū zhě, qì zhī shǐ yě [wèi jiàn (wùzhì zhī) xíng]; tài shǐ zhě, xíng zhī shǐ yě [wèi jiàn (wùzhì zhī) zhì]; tài sù zhě, zhì zhī shǐ yě [wèi jiàn (wùzhì zhī) tǐ]
	Translation	Taiyi (the great easiness), at which *qi* is too weak to be seen (though it is produced inside the nebulous void); taichu (the great origin), at which *qi* emerges [yet, without xíng, that is, (matter's) shape]; taishi (the great inaugural), at which xíng emerges [from *qi*, yet, without zhì, that is, (matter's) quality]; taisu (the great simplicity), at which zhì emerges [from *qi* and xíng, yet, without tǐ, that is, (matter's) entity]
Page 44	**Phrase**	气形质具而未相离，故曰浑沦。浑沦者，言万物相浑沦而未相离也；视之不见，听之不闻，循之不得，故曰易也；易无形埒，易变而为一 . . . 一者，形变之始也
	Pinyin	qì xíng zhì jù ér wèi xiāng lí, gù yuē hún lún. hún lún zhě, yán wàn wù xiāng hún lún ér wèi xiāng lí yě. shì zhī bù jiàn, tīng zhī bù wén, xún zhī bù dé, gù yuē yì yě; yì wú xíng liè, yì biàn ér wéi yī . . . yī zhě, xíng biàn zhī shǐ yě
	Translation	The three ingredients of *qi*, xíng (shape), zhì (quality) are blended in a state of hundun (chaos), where nothing can be separated from anything else; (this is a state) invisible to sight, inaudible to hearing, and intangible to touching, thus recognized as *yi*; it possesses neither shape nor bounds and expresses itself as the oneness of all . . . this oneness is the beginning of the following unprecedented deformations (i.e., the distortions or alterations in shape or structure)
Page 44	**Phrase**	视而不见，名曰夷；听之不闻，名曰希；搏之不得，名曰微；此三者不可致诘，故混而为一。 其上不徼，其下不昧，绳绳兮不可名，复归于无物。 是谓无状之状，无物之象，是谓惚恍；迎之不见其首，随之不见其后 . . . 能知古始，是谓道纪

TABLE C.6. CLASSICAL CHINESE PHRASES (*cont.*)

Page		
Page 44	Pinyin	shì ér bù jiàn, míng yuē yí; tīng zhī bù wén, míng yuē xī; bó zhī bù dé, míng yuē wēi cǐ sān zhě bù kě zhì jié, gù hùn ér wéi yī qí shàng bù jiào, qí xià bù mèi, shéng shēng xī bù kě míng, fù guī yú wú wù shì wèi wú zhuàng zhī zhuàng, wú wù zhī xiàng, shì wèi hū huǎng. yíng zhī bù jiàn q shǒu, suí zhī bù jiàn qí hòu . . . néng zhī gǔ shǐ, shì wèi dào j
	Translation	Invisible to seeing, thus called the colorless, *yí*; inaudible to hearing, thus called the soundless, xī; intangible to touching, thus called the shapeless, wēi; which are so inseparable, in no way to be defined independently, that they exist as a mixed-up unity (to express all the patterns of wuji). Outside the unity, there is no more lightness; and, inside the unity, there is no more darkness; thus, it is vague enough to defy any description and can be recategorized as nothingness. For this reason, the unity is a form without form, an image of matter without matter, hence it is called huhuang (seemingly visible but invisible), of which neither can the front be seen when facing it nor the back be seen when following it . . . It is the demonstration of *dao*, from which the beginning of the past can be known
Pages 45–46	Phrase	动而无静，静而无动，物(性)也；动而无动，静而无静，神(性)也；阴阳不测之谓神(性)
	Pinyin	dòng ér wú jìng, jìng ér wú dòng, wù (xìng) yě; dòng ér wú dòng, jìng ér wú jìng, shén (xìng) yě; yīn yáng bù cè zhī wèi shén (xìng)
	Translation	Either being *yang* in activity instead of *yin* in tranquility, or being *yin* instead of *yang*, forms wùxìng (the thingness of things); either being active but with a lack of *yang*, or being tranquil but with a lack of *yin*, forms shén xìng (the unfathomability of things); the unpredictability of *yin* and *yang* is called unfathomability
Page 46	Phrase	动而无动，静而无静，非不动不静也；动中有静，静中有动 . . . 道具于阴而行于阳
	Pinyin	dòng ér wú dòng, jìng ér wú jìng, fēi bù dòng bù jìng yě; dòng zhōng yǒu jìng, jìng zhōng yǒu dòng; dào jù yú yīn ér xíng yú yáng
	Translation	Both being active but lacking *yang* and being tranquil but lacking *yin* do not mean being fully devoid of *yang* and *yin*, respectively; there always exist tranquility within activeness and activeness within tranquility . . . *dao* resides in *yin* and acts in *yang*
Page 47	Phrase	生生之谓易
	Pinyin	shēng shēng zhī wèi yì
	Translation	(The process of) endless alternative interchanging and generation (of *yin* and *yang*) is what is called *yi*
Page 47	Phrase	继之者 . . . 谓化育之功，阳之事也；成之者 . . . 谓物之所受 . . . 阴之事也
	Pinyin	jì zhī zhě . . . wèi huà yù zhī gōng, yáng zhī shì yě; chéng zhī zhě . . . wèi wù zhī suǒ shòu . . . yīn zhī shì yě

TABLE C.6. CLASSICAL CHINESE PHRASES (*cont.*)

Page		
Page 47	Translation	To sustain *dao* . . . is the duty of *yang* in the accomplishment of transformation and cultivation, while to achieve *dao* . . . is the duty of *yin* . . . in the nourishment of the promotion and the diversity
Page 47	Phrase	生生之谓易；阴生阳，阳生阴，其变无穷
	Pinyin	shēng shēng zhī wèi yì; yīn shēng yáng, yáng shēng yīn, qí biàn wú qióng
	Translation	Endless alternative interchanging and generation is called *yi*; *yin* generates *yang* and *yang* generates *yin* in myriad transformations
Page 47	Phrase	有之以为利，无之以为用
	Pinyin	yǒu zhī yǐ wéi lì, wú zhī yǐ wéi yòng
	Translation	When *yin* (or *yang*) dominates, it serves as the functioning polarity of the transformation; and, when it does not dominate, it provides the condition for the other to behave as the functioning polarity of the transformation
Page 47	Phrase	两仪生四象；四象生八卦；刚柔相摩，八卦相荡
	Pinyin	liǎng yí shēng sì xiàng; sì xiàng shēng bā guà; gāng róu xiāng mó, bā guà xiāng dàng
	Translation	Two polarities produce four digrams; four digrams produces eight trigrams; mutual communion and alternation (happen) between the resolute (*yang*) and the yielding (*yin*), together with the full combinations of the two eight-trigram sets
Page 72	Phrase	开物；闭物
	Pinyin	kāi wù; bì wù
	Translation	All things begin to start; all things terminate
Page 72	Phrase	致虚极，守静笃，万物并作，. . . 各复归其根；归根曰静，是为复命。复命曰常，知常曰明
	Pinyin	zhì xū jí, shǒu jìng dǔ, wàn wù bìng zuò, . . . gè fù guī qí gēn; guī gēn yuē jìng, shì wéi fù mìng. fù mìng yuē cháng, zhī cháng yuē míng
	Translation	Brought to an ultimate at which an unwavering stillness is kept; From this, all things come into being, and go through similar processes of evolution; Then, . . . they respectively return to the same root; The root-returning is called tranquilization, that is, the return to destiny; The fulfillment of the life-resumption is the unchanging law of regularity, the consciousness of which is called being enlightened
Page 124	Phrase	乾坤离坎四正卦 . . . 用之以作闰卦也
	Pinyin	qián kūn lí kǎn sì zhèng guà . . . yòng zhī yǐ zuò rùn guà yě
	Translation	The four principal hexagrams, Qian, Kun, Kan, Li . . . are used for the intercalations of the timekeeping

References

Abe-Ouchi A., F. Saito, K. Kawamura, et al. 2013. "Insolation-Driven 100,000-Year Glacial Cycles and Hysteresis of Ice-Sheet Volume." *Nature* 500, 190–93.

Abramowicz, M. 1990. "Centrifugal-Force Reversal near a Schwarzschild Black Hole." *Monthly Notices of the Royal Astronomical Society* 245 (4): 720–28.

Adams F. C., and G. Laughlin. 1997. "A Dying Universe: The Long-Term Fate and Evolution of Astrophysical Objects." *Reviews of Modern Physics* 69 (2): 337–72.

Amberg, A. 2011. "Incarnational Spirituality in Partnership with the Powers of the Universe." Master's thesis, Lorian Center for Incarnational Spirituality, Issaquah, WA.

Andina, T. 2014. *Bridging the Analytical Continental Divide: A Companion to Contemporary Western Philosophy*. Leiden, Netherlands: Brill.

Andreeva, A., and D. Steavu. 2016. *Transforming the Void: Embryological Discourse and Reproductive Imagery in East Asian Religions*. Leiden, Netherlands: Brill.

Angelo, J. A. 2006. *Encyclopedia of Space and Astronomy*. New York: Facts on File.

Anonymous (700–750). The yellow emperor's classic of the secret talisman (J. Legge trans.). Chinese Text Project website.

Anonymous. Tang dynasty. Book of the eight effulgences of the primordial chaos. Chinese Text Project website.

Antinoff, S. 2010. *Spiritual Atheism*. Berkeley, CA: Counterpoint Press.

Ashton, T. 1948. *The Industrial Revolution 1760–1830*. 4th ed. Oxford: Oxford University Press.

Baker, C. 2015. *Love in the Age of Ecological Apocalypse: Cultivating the Relationships We Need to Thrive*. Berkeley, CA: North Atlantic Books.

Barnard, L., and T. B. Hodges. 1958. *Readings in European history*. New York: Macmillan.

Bars, I., and J. Terning. 2009. *Extra Dimensions in Space and Time*. New York: Springer.

BEC CREW. 2015. "This Timeline Shows the Entire History of the Universe, and Where It's Headed." Science Alert website.

Beekes, R. S. P. 2009. *Etymological Dictionary of Greek*. Leiden, Netherlands: Brill.

Belenkiy, A. 2012. "Alexander Friedmann and the Origins of Modern Cosmology." *Physics Today* 65 (10): 38–43.

Bidlack, B. B. 2010. "Teilhard de Chardin in China: Challenge and Promise." *China Heritage Quarterly* 23, online.

Birx, H. J. 1999. "The Phenomenon of Pierre Teilhard de Chardin." *Religious Humanism* 33 (1), online.

———. 2015. "Pierre Teilhard de Chardin: Critical Reflections." *Anthropologia Integra* 6 (1): 7–22.

Bod, R., and J. Kursell. 2015. "Introduction: The Humanities and the Sciences." *Isis* 106 (2): 337–40.

Bohmadi-Lopez, M., P. F. Gonzalez-Diaz, and P. Martin-Moruno. 2008. "Worse Than a Big Rip?" *Physical Letters B* 659 (1–2): 1–5.

Bondi, H. 1952. *Cosmology*. Cambridge: Cambridge University Press.

Bridle, S. 2016. "The Divinization of the Cosmos: An Interview with Brian Swimme on Pierre Teilhard de Chardin."

Bull, P., Y. Akrami, J. Adam, et al. 2016. "Beyond Λ CDM: Problems, Solutions, and the Road Ahead." *Physics of the Dark Universe* 12, 56–99.

Buswell, R. E., and D. S. Lopez. 2014. *The Princeton Dictionary of Buddhism*. Princeton, NJ: Princeton University Press.

Cai, S. Song dynasty. Supreme principles in great plan, inner chapter, volume 2. Chinese Text Project website. Numerological group, 7th Zi branch. Annotated Catalog of the Complete Imperial Library website.

Caldwell, R. R., M. Kamionkowski, and N. N. Weinberg. 2003. "Phantom Energy and Cosmic Doomsday." *Physical Review Letters* 91 (7): 1–4.

Campisi. P., D. La Rocca, and G. Scarano. 2012. "EEG for Automatic Person Recognition." *Computer* 45, 87–89.

Canale, F. 2005. "The Quest for the Biblical Ontological Ground of Christian Theology. *Journal of the Adventist Theological Society* 16 (1–2): 1–20.

Cañizares, J. S. 2011. "Review of the Cycles of Time: An Extraordinary New Vision of the Universe." *Yearbook Filosófic* 44 (2): 416–18.

Carroll, S. M. 2005. "Why (Almost All) Cosmologists Are Atheists. *Faith and Philosophy* 22 (622): 1–13, online.

Cartwright, M. 2015. "Brahma." Ancient History Encyclopedia website.

Castillo, M. 2012. "The Omega Point and Beyond: The Singularity Event." *American Journal of Neuroradiology* 33 (3): 393–95.

Cataldo, C. 2016a. "Further Remarks on the Oscillating Universe: An Explicative Approach." *Research and Reviews: Journal of Pure and Applied Physics* 4 (3): 33–37.

———. 2016b. "A Simplified Model of Oscillating Universe: Alternative Deduction of Friedmann Lemaître Equations with a Negative Cosmological Constant." *Research and Reviews: Journal of Pure and Applied Physics* 4 (2): 1–3.

Caughey, E. M. 2016. "Hawking Radiation Screening and Penrose Process Shielding in the Kerr Black Hole." *The European Physical Journal C* 76 (179): 1–11.

Cavedon, M. 2013. "Independent Science, Integrated Theology: How Process Theology Can Inform Christian Orthodox Theodicy." *SSRN* 1–9. SSRN website.

Center for the Story of the Universe, 2004. "Announcing a New DVD Series from Cosmologist, Brian Swimme: Powers of the Universe." *Gatherings: Journal of the International Community for Ecopsychology.*

Chambers, M., B. Hanawalt, T. K. Rabb, et al. 2007. *The Western Experience.* 9th ed. New York: McGraw-Hill.

Chan, W.-T. 1969. *A Source Book in Chinese Philosophy.* Princeton, NJ: Princeton University Press.

Chao, J. 1989. *China Mission Handbook.* Hong Kong: Chinese Church Research Center.

Chen, C.-M. 2011. On the activity of the principles (li) in Chu Hsi's interpretation of t'ung-shu and an explanation of the diagram of the great ultimate. *Zhongzheng University Journal of Chinese Language Studies* 18 no. 2: 1–28.

Chen, D.-S. 2015. Chapter 5 of treatise of remarks on the trigrams. In: Extensive interpretation of the original Zhou-I: Interpretation by experts over one hundred years. Chengdu, China: Bashu Press.

Chen, M.-D. 1994. Theory of cosmological expansion in ancient China. *Journal of Natural Science History Research* 1, 27–31.

Chen, Q.-Y. 2002. New annotation of proofreading texts on Lushi Chunqiu. Shanghai, China: Shanghai Ancient Rare Books Publishing House.

Cheng, C.-Y. 2019. "Consciousness: Chinese Thought." In M. C. Horowitz,

New Dictionary of the History of Ideas. New York: Charles Scribner's Sons.

Cheng, Z. 2006. The five meanings of "change" and onto-hermeneutics of I Ching. *Journal of Chinese Language and Literature of National Taipei University* 1, 1–32.

Clayton, L., J. W. Attig, D. M. Mickelson, et al. 2006. *Glaciation of Wisconsin*. 3rd ed. Madison: Wisconsin Geological and Natural History Survey, Educational Series 36.

Clayton, P. 2010. "Panentheisms East and West." *Sophia* 49 (2): 183–91.

CMS (Compact Muon Solenoid). 2011. "Story of the Universe: From the Big Bang to Today's Universe." CMS website.

Cobb, J. B., and D. R. Griffin. 1976. *Process Theology: An Introductory Exposition*. Philadelphia: The Westminster Press.

Coffey, P. 1918. *Ontology, or The Theory of Being: An Introduction to General Metaphysics*. London, England: Longmans, Green and Co. Google Books version.

Cooper, J. W. 2006. "Teilhard de Chardin's Christocentric Panentheism." In *Panentheism, the Other God of the Philosophers: From Plato to the Present*. Grand Rapids, MI: Baker Academic.

Copernicus, N. 1995. *On the Revolutions of Heavenly Spheres*. Trans. C. G. Wallis. New York: Prometheus Books.

Cornford, F. M. 1997. *Plato's Cosmology: The Timaeus of Plato*. Indianapolis, IN: Hackett Publishing.

CSIRO. n.d. "The Big Bang and the Standard Model of the Universe." Australia Telescope National Facility/Commonwealth Scientific and Industrial Research Organisation website.

Culp, J. 2021. "Panentheism." *The Stanford Encyclopedia of Philosophy* website.

Damdul, G. D. 2019. "Ontological Reality: Quantum Theory and Emptiness in Buddhist Philosophy." In S. R. Bhatt, *Quantum Reality and Theory of Śūnya*, 345–50. Singapore: Springer.

Deng, Z.-X. Qing dynasty. Book two of the treatise of supreme world-ordering principles. Chinese Text Project website. Volume 8 of the great compendium on nature and principle, Zi branch, Annotated Catalog of the Complete Imperial Library.

Digges, T. 1999 [1576]. "A Perfect Description of the Celestial Orbs." In *As the World Turned: A Reader on the Progress of the Heliocentric Argument from Copernicus to Galileo*. Hanover, NH: Mathematics Across the Curriculum Project, Dartmouth College.

———. 2001 [1576]. "A Perfect Description of the Celestial Orbs. In K. Aughterson, *The English Renaissance: An Anthology of Sources and Documents*, 355–57. London: Routledge.

Ding, Z. 2005. "The Numerical Mysticism of Shao Yong and Pythagoras." *Journal of Chinese Philosophy* 32 (4): 615–32.

Dorling Staff. 1994. *The Visual Dictionary of Ancient Civilizations*. New York: Dorling Kindersley Publishing.

Dong, Z.-S. 206 BCE. Luxuriant gems of the spring and autumn. Chinese Text Project website.

Drasny, J. 2011. *The Yi-Globe: The Image of the Cosmos in the "Yijing."* 2nd ed. Available online.

Dusek, V. 1999. *The Holistic Inspirations of Physics: The Underground History of Electromagnetic Theory*. New Brunswick, NJ: Rutgers University Press.

Einstein, A. 1916. "The Foundation of the General Theory of Relativity. *Annalen der Physik* 354 (7): 769–822.

———. 1932. "Prologue." In M. Planck, *Where Is Science Going? The Universe in the Light of Modern Physics*. New York: W. W. Norton and Company.

———. 1934. "On the Method of Theoretical Physics." In *Mein Weltbild*. Amsterdam: Querida Verlag.

———. 1997 [1915]. "The Field Equations of Gravitation." In *The Collected Papers of Albert Einstein*. Vol. 6, *The Berlin Years, Writings 1914–1917*. Translated by A. Engel, 117–20. Princeton University Press website.

Falkus, M. 1987. *Britain Transformed: An Economic and Social History 1700–1914*. 4th ed. Ormskirk, England: Causeway Press.

Farnes, J. S. 2018. "A Unifying Theory of Dark Energy and Dark Matter: Negative Masses and Matter Creation within a Modified LambdaCDM Framework." *Astronomy and Astrophysics* 620, article A92.

Feng, Y.-L. 2004. A short history of Chinese philosophy (F. S. Zhao trans.). Beijing: New World Press. Available in English by Fung, Y.-L. 1997. *A Short History of Chinese Philosophy*. New York: Free Press.

Frautschi, S. C., R. P. Olenick, T. M. Apostol, et al. 2007. *The Mechanical Universe: Mechanics and Heat*. adv ed. Cambridge: Cambridge University Press.

Friedman, A. 1922. "Über die Krummung des Raumes." *Zeitschrift fur Physik* 10, 377–86.

Frolov, V. P., and I. D. Novikov. 2012. *Black Hole Physics: Basic Concepts and New Developments*. Berlin: Springer Science and Business Media.

Fuentes, C. S.-P. 1996. *Panentheism in Teilhard de Chardin: A Creative Synthesis*. Boston: Boston University Press.

Fung, Y. L. 1948. *A Short History of Chinese Philosophy*. Edited and translated by D. Bodde. New York: Macmillian.

Galilei, G. 1610. *Sidereus nuncius* [Starry messenger]. Translated by P. Barker. Oklahoma City, OK: Byzantium Press. See also Sidereus Nuncius [The Sidereal Messenger]. Translated by Albert Van Helden. Chicago: University of Chicago Press, 1989.

———. 1914 [1638]. *Dialog Concerning Two New Sciences*. Translated by H. Crew and A. de Salvio. New York: Macmillan.

Ge, J. 2014. "The Tradition Wisdom of the Chinese Calendar and Urban Development in Ancient China. In *Traditional Wisdom and Modern Knowledge for the Earth's Future*, edited by K. Okamoto and Y. Ishikawa. Lectures given at the plenary sessions of the International Geographical Union Kyoto Regional Conference. Tokyo, Japan: Springer, 34–50.

Gernet, J. 1996. *A History of Chinese Civilization*. Translated by J. H. Foster and C. Hartman. 2nd ed. Cambridge: Cambridge University Press.

Gethin, R. 1998. *The Foundations of Buddhism*. Oxford: Oxford University Press.

Ghislain, D. 2009. *Panmobilism and Optimism in Teilhardian Humanism: 2.3.1. Cosmogenesis and Christogenesis*. Memoir online website.

Giddens, A. 1974. *Positivism and Sociology*. London: Pearson Education.

Glattfelder, J. B. 2019. "Ontological Enigmas: What Is the True Nature of Reality?" In *Information-Consciousness-Reality*. The Frontiers Collection. Cham, Switzerland: Springer.

Göcke, B. P. 2013. "Panentheism and Classical Theism." *SOPHIA* 52 (1): 61–75.

Gregersen, N. H. 2004. "Three Varieties of Panentheism." In *In Whom We Live and Move and Have Our Being: Panentheistic Reflections on God's Presence in a Scientific World*, edited by P. Clayton and A. Peacocke. Grand Rapids, MI: William B. Eerdmans, 19–35.

Grim, J. 2016. "President's Corner." *Teilhard Perspective* 49 (1): 1–16.

Grim, J., and M. E. Tucker. 2017. "History." Yale Forum on Religion and Ecology website.

Gurzadyan, V. G., and R. Penrose. 2010. "Concentric Circles in WMAP Data May Provide Evidence of Violent Pre-Big-Bang Activity." Available on the Archive of Cornell University website (arXiv).

Hakak, S. M. 2002. "Prime Mover (Primum Mobile) as Viewed by Aristotle." *Maneh-Ye-Mofid* 8 (2): 115–24.

Hall, J. M. 2007. *A History of the Archaic Greek World, ca. 1200–479 BCE.* Malden, MA: John Wiley and Sons.

Halpern, P., and N. Tomasello. 2016. "Size of the Observable Universe." *Advances in Astronomy* 1 (3): 135–37.

Halyo, E., B. Kol, A. Rajaraman, et al. 1997. "Counting Schwarzschild and Charged Black Holes." *Physics Letters B* 401 (1–2): 15–20.

Hameroff, S., and R. Penrose. 2003. "Conscious Events as Orchestrated Space-Time Selections." *Neuro Quantology* 1 (1): 10–35.

———. 2014. "Consciousness in the Universe: A Review of the 'Orch OR' Theory." *Physics of Life Reviews* 11 (1): 39–78.

Hardy, J. M. 1998. "Influential Western Interpretations of the *Tao-Te-Ching.*" In *Lao-tzu and the "Tao-Te-Ching": Studies in Ethics, Law, and the Human Ideal,* edited by L. Kohn and M. LaFargue, 165–88 Albany: State University of New York Press.

Hartle, J. B., and S. W. Hawking. 1983. "The Wave Function of the Universe." *Physical Review D* 28 (12): 2960–75.

Hartle, J. B., S. W. Hawking. and T. Hertog. 2008. "Classical Universes of the No-Boundary Quantum State." *Physical Review D* 77 (12): 123537.

Hawking, S. W. 1975. "Particle Creation by Black Holes." *Communications in Mathematical Physics* 43 (3): 199–220.

———. 1988. *A Brief History of Time.* New York: Bantam.

———. 1996. *The Beginning of Time.* Academic lecture available on the hawking.org.uk website.

Hawking, S. W., and T. Hertog. 2018. "A Smooth Exit from Eternal Inflation?" *Journal of High Energy Physics,* 147, 1–13.

Hawking, S. W., and R. Penrose. 1970. "The Singularities of Gravitational Collapse and Cosmology." *Proceedings of the Royal Society A: Mathematical, Physical and Engineering Sciences* 314 (1519): 529–48.

He, N. 1998. Volume 3: Patterns of heaven. In: *New collections of ancient Chinese philosophers: Collected interpretations of Huai-Nan-Zi.* Beijing, China: Zhonghua Book Company.

Hegel, G. W. F. 1892. *Lectures on the History of Philosophy.* Vol. 1. Translated to English by E. S. Haldane. London: Kegan Paul, Trench, Trübner and Co.

———. 1983. *Lectures on the history of philosophy.* Translated to Chinese by L. He, T. Q. Wang. Beijing: Commercial Press. Note that this version has more of Hegel's lecture notes compared with the English version, particularly those related to Chinese works.

———. 2011. *Lectures on the Philosophy of World History* Vol.1, *Manuscripts of the Introduction and the Lectures of 1822–3*. Translated and edited by R. F. Brown and P. C. Hodgson. Oxford: Clarendon Press.

Hetherington, N. S. 2014. *Encyclopedia of Cosmology: Historical, Philosophical, and Scientific Foundations of Modern Cosmology*. Routledge Revivals edition. New York: Routledge.

Hinton, D., trans. 2013. *The Four Chinese Classics: "Tao Te Ching," "Analects," "Chuang Tzu," "Mencius."* Berkeley, CA: Counterpoint Press.

Hitchcock, B. 2013. *The Evolution of Unitarianism and Universalism: A Liberal Inheritance*. High Plains Church Unitarian Universalist website.

Hohenkerk, C. Y., B. D. Yallop, C. A. Smith, et al. 1992. "Celestial Reference Systems." In *Explanatory Supplement to the Astronomical Almanac*, edited by P. K. Seidelmann, 95–198. Mill Valley, CA: University Science Books.

Holton, G. J., and S. G. Brush. 2001. *Physics, the Human Adventure: From Copernicus to Einstein and Beyond*. 3rd ed. Piscataway, NJ: Rutgers University Press.

Hu, Y., and C. D. Li . 2004. "Had Leibniz not seen the *Xian tian tu* before he founded the binary system?—A textual study of European literature about culture exchange between China and the West in the seventeenth century." *Research of the Yijing* 2, 66–71.

———. 2006. *Investigation of Leibniz's binary system and Fuxi's diagram of the eight trigrams*. Shanghai: Shanghai People Press.

Huang, S.-C. 1999. *Essentials of Neo-Confucianism: Eight Major Philosophers of the Song and Ming Periods*. Westport, CT: Greenwood Publishing.

Hubble, E. 1929. "A Relation between Distance and Radial Velocity among Extra-Galactic Nebulae." *Proceedings of the National Academy of Sciences* 15 (3): 168–73.

Jaroszkiewicz, G. 2016. *Images of Time: Mind, Science, Reality*. Oxford: Oxford University Press.

Jiang, S. and W. X. Tang. 2002. "Daoist science and technology in Chinese history." In: *Volume of Han, Wei, and two-Ji Dynasties*. Beijing, China: Science Press.

Jibu, M., and K. Yasue. 1995. *Quantum Brain Dynamics and Consciousness: An Introduction*. Amsterdam, Netherlands: John Benjamins.

Jiménez, J. B., R. Lazkoz, D. Sáez-Gómez, et al. 2016. "Observational Constraints on Cosmological Future Singularities." *The European Physical Journal C* 76 (631): 1–13.

Jun, W. 2014. *Ancient Chinese Encyclopedia of Technology*. New York: Routledge.

Jung, C. G. 1947. Foreword to *The Secret of the Golden Flower: A Chinese Book of Life*. Translated by R. Wilhelm and C. F. Baynes, xiii–xiv. London: Kegan Paul, Trench, Trübner and Co. First published in 1930.

Kant, I. 1872. *Critique of Pure Reason*. Translated by J. M. D. Meiklejorn. London: Bell and Daldy.

———. 2008. *Universal Natural History and Theory of the Heavens*. Translated by I. Johnston. Arlington, VA: Richer Resources Publications.

Kant, I. 2016. *The Critique of Pure Reason*. Translated by J. M. D. Meiklejorn. Woodstock, ON: Devoted Publishing.

Kazlev, M. A. 2000. *The Big Bang and the History and Evolution of the Universe*. The Khepher website.

Kelly, S. 2015. "Cosmological Wisdom and Planetary Madness." *Tikkun* website.

Kelvin, L., and P. G. Tait. 1912. *Treatise on Natural Philosophy, Part I*. Cambridge: Cambridge University Press.

Khoo, F. S., and Y. C. Ong. 2016. "Lux in Obscuro: Photon Orbits of Extremal Black Holes Revisited." *Classical and Quantum Gravity* 33 (23): 235002.

———. 2017. "Corrigendum: Lux in Obscuro: Photon Orbits of Extremal Black Holes Revisited (2016 *Class. Quantum Grav.* 33 235002)." *Classical and Quantum Gravity* 34 (21): 219501.

Kim, Y. K. 1978. "Hegel's Criticism of Chinese Philosophy." *Philosophy East and West* 28 (2): 173–80.

Kisak, P. F. 2015. *A Chronology of Our Universe: A Timeline of Cosmological and Astronomical Events*. Scotts Valley, CA: CreateSpace Independent Publishing.

Kragh, H. 1999. *Cosmology and Controversy*. Princeton, NJ: Princeton University Press.

———. 2015. "The 'New Physics.'" In *The Fin-de-siècle World*, edited by M. Saler, 441–55. Abingdon, England: Routledge.

Kreis, S. 2006. "Lecture 3: Egyptian Civilization." The History Guide website.

———. 2009. "Lecture 5: Homer and the Greek Renaissance, 900–600 BC." The History Guide website.

———. 2014. "Lecture 2: Ancient Western Asia and the Civilization of Mesopotamia." The History Guide website.

Kutner, M. 2003. *Astronomy: A Physical Perspective*. Cambridge: Cambridge University Press.

Landow, G. P. 2012. "The Industrial Revolution: A Timeline." The Victorian Web website.

Lane, D. H. 1996. *The Phenomenon of Teilhard: Prophet for a New Age*. Macon, GA: Mercer University Press.

Laozi (ca. 475–471 BCE). Daodejing/Tao-Teh-Ching, also known as Lao-Tzu. Chinese Text Project website.

Laplace, P.-S. 1829. *Traités de Mécanique céleste*. Translated by N. Bowditch as *Treatise on Celestial Mechanics*. 4 vols. Boston: The Press of Isaac R. Butts. Published in 1969 as *Celestial Mechanics, Vol. 5*. Edited by N. Bowditch. Bronx, NY: Chelsea Publishing.

Laplace, P.-S. 1830. *Exposition du système du monde*. Translated by Rev. H. Harte as *The system of the world*. Dublin, Ireland: University Press.

———. 1799. "Beweis des satzes, dass die anziehende kraft bey einem welt-körper so gross seyn könne, dass das licht davon nicht ausströmen kann." *Allgemeine Geographische Ephemeriden* 4, 1–6.

Lefferts, M. 2014. "Fractal Holographic Synergetic Universe." Cosmometry Project website.

Legge, J. 1891. *The Sacred Books of China: The Writings of Kwang-sze, Part II*. In *The Sacred Books of the East*, vol. 40, edited by F. M. Müller, 1–57. Oxford: Clarendon Press.

———. 1963. *The I Ching: The Book of Changes*. 2nd ed. In *The Sacred Books of the East*, vol. 16, edited by F. M. Müller, 57–207. New York: Dover Publications.

Le Grice, K. 2011. *The Archetypal Cosmos: Rediscovering the Gods in Myth, Science and Astrology*. Edinburgh: Floris Books.

Lemaître, G. 1927. "Un univers homogène de masse constant et de rayon croissant rendant compte de la vitesse radiale des nébuleuses extragalactiques." *Annales de la Société Scientifique de Bruxelles*, A47, 49–59. Published in 1931 as "A Homogeneous Universe of Constant Mass and Growing Radius Accounting for the Radial Velocity of Extragalactic Nebulae." *Monthly Notices of the Royal Astronomical Society* 91, 483–90.

Leplin, J. 1997. *A Novel Defense of Scientific Realism*. Oxford: Oxford University Press.

Lessner, G. 2006. "Oscillating Universe—An Alternative Approach in Cosmology." *Astrophysics and Space Science* 306 (4): 249–57.

———. 2011. "Oscillating Universe." *Journal of Modern Physics* 2, 1099–103.

Lestone, J. P. 2007. "A Possible Connection between Hawking Radiation and the Electric Charges of Fundamental Particles." CERN website.

Levit, G. S. 2000. "The Biosphere and the Noosphere Theories of V. I. Vernadsky and P. Teilhard de Chardin: A Methodological Essay." *International Archives on the History of Science/Archives Internationales D'Histoire des Sciences* 50 (144): 160–76.

Li, X. P. 2002. A tentative study of Meng Xi and Jing Fang's gua-qi theory. *Zhou-I Research* 3, 66–72.

Li, Y. S. 1983. "Discussions of a few issues related to yin-fu jing: The classic of the secret talisman." *China Daoism* 1, 1–21.

Liang, W. X. 2007. *Research on hexagrams' vital power of Yi-Jing in Han dynasty.* Jinan: Qilu Book Publishing Co., Ltd.

Liezi. 475–221 BCE. *Lie Zi: Heaven's gifts.* Chinese Text Project website.

Linton, C. M. 2004. *From Eudoxus to Einstein: A History of Mathematical Astronomy.* Cambridge: Cambridge University Press.

Liu, An. 139 BCE. *Huainanzi.* Chinese Text Project website.

———. 2010. *The* Huainanzi: *A Guide to the Theory and Practice of Government in Early Han China.* Translated by J. S. Major, S. Queen, A. Meyer, et al. New York: Columbia University Press.

Liu, C. L. 2013. "On the philosophy and cultural connotations of gua qi in Han dynasty: Case study about Meng Xi." *Qilu Journal* 5, 20–24.

Liu, Z., and L. Liu. 2010. *Essentials of Chinese Medicine.* Vol. 1, *Foundations of Chinese Medicine.* Dordrecht, Germany: Springer.

Lochner, J. 1998. "Ask an Astrophysicist" answer. NASA Goddard Space Flight Center website.

Lü, B. W. 2001. *The Annals of Lü Buwei.* Translated by J. Knoblock and J. Riegel. Palo Alto, CA: Stanford University Press.

Luquet, W. 2006. "Union Differentiates: Pierre Teilhard de Chardin's Philosophy Applied to Couple Relationships." *The Family Journal* 14 (2): 144–50.

Ma, Z. 2016. "Comparison between Hegel's Being-Nothing-Becoming and Yi-Jing's Yin-Yang-I (Change)." *Asian Research Journal of Arts and Social Sciences* 1 (6): 1–15.

———. 2017. "Plasma Brain Dynamics (PBD): A Mechanism for EEG Waves under Human Consciousness." *Cosmos and History: The Journal of Natural and Social Philosophy* 13 (2): 185–203.

———. 2018a. "Plasma Brain Dynamics (PBD): II. Quantum Effects on Consciousness." *Cosmos and History: The Journal of Natural and Social Philosophy* 14 (1): 91–104.

———. 2018b. "Review of *The Powers of the Universe* by Brian Swimme." DVD review. *Asian Research Journal of Arts and Social Sciences* 5 (2): 1–8.

Ma, Z. G., and H. C. Zeng. 2024. *The Tao of Cosmos (III): Holographic Four Pillars (Ba Zi) in Accordance with Shao Yong's World-Ordering Principles.* eBook, Kindle edition.

Mabkhout, S. A. 2015. *The Hyperbolic Universe.* Edited by A. Gesica. Saarbrücken, Germany: Lambert Academic.

MacDougal, D. W, 2012. *Newton's Gravity: An Introductory Guide to the Mechanics of the Universe.* New York: Springer.

Maciocia, G. 2015. *The Foundations of Chinese Medicine.* 3rd ed. Edinburgh: Elsevier.

Maeder, A. 2017. "Dynamical Effects of the Scale Invariance of the Empty Space: The Fall of Dark Matter?" *The Astrophysical Journal* 849 (2): 158.

Major, J. S. 1993. *Heaven and Earth in Early Han Thought: Chapters Three, Four, and Five of the* Huainanzi. Albany: State University of New York Press.

Makeham, J. 2010. *Dao Companion to Neo-Confucian Philosophy.* Dordrecht, Germany: Springer.

Manfredi, G. 2005. "How to Model Quantum Plasmas." *Fields Institute Communications* 46, 263–87.

Mark, E. 2016. "Shang Dynasty." Ancient History Encyclopedia website.

Martini, M. 1658. *Sinicae Historiae Decas Prima.* Original from the Bavarian State Library. Google Books.

Mastin, L. 2010. "Neurons and Synapses." The Human Memory website.

———. 2019. "Timeline of the Big Bang." Physics of the Universe website.

Mathematics and Nature. 2017. "Gif (1), (2), (3): Lissajous Figure as Pattern on Horn Torus." Science and Math website.

Matthews, M. R. 2017. *History, Philosophy and Science Teaching: New Perspectives.* Cham, Switzerland: Springer.

Mazlish, B. 2004. *Civilization and Its Contents.* Palo Alto, CA: Stanford University Press.

McCluskey, S. C. 2000. *Astronomies and Cultures in Early Medieval Europe.* Cambridge: Cambridge University Press.

McGrath, A. E. 2006. *Scientific Theology.* Vol. 1, *Nature.* New York: T and T Clark.

Mei, G. C. 1739. *Imperially endorsed treatise on harmonizing times and distinguishing directions.* Chinese Text Project website.

Melia, F. 2013. "The $R_H = ct$ Universe without Inflation." *Astronomy and Astrophysics* 553 (A76): 1–6.

———. 2015. "The Cosmic Equation of State." *Astrophysics and Space Science* 356 (2): 393–98.

Mesle, C. R. 1993. *Process Theology: A Basic Introduction*. St. Louis: Chalice Press.

Mickey, S. 2016. *Whole Earth Thinking and Planetary Coexistence: Ecological Wisdom at the Intersection of Religion, Ecology, and Philosophy*. Abingdon, England: Routledge.

Mickey, S., S. Kelly, and A. Robbert. 2017. *The Variety of Integral Ecologies: Nature, Culture, and Knowledge in the Planetary Era*. Albany: State University of New York Press.

Midbon, M. 2000. "A Day without Yesterday: Georges Lemaître and the Big Bang." *Commonweal* 127 (6): 18–19.

Midgley, M. 2007. *Earthy Realism: The Meaning of Gaia*. Exeter, England: Imprint Academic.

Mikkelsen, G. 2006. Review of *A Study of the History of Nestorian Christianity in China and Its Literature in Chinese: Together with a new English Translation of the Dunhuang Nestorian Documents* by Li Tang. *China Review International* 14 (1): 232–35.

Misner, C. W., K. S. Thorne, and J. A. Wheeler. 1973. *Gravitation*. San Francisco, CA: W. H. Freeman.

Montgomery, C., W. Orchiston, and I. Whittingham. 2009. "Michell, Laplace and the Origin of the Black Hole Concept." *Journal of Astronomical History and Heritage* 12 (2): 90–96.

Morin, E., and A. B. Kern. 1999. *Homeland Earth: A Manifesto for the New Millennium*. Translated by S. Kelly and R. Lapointe. Cresskill, NJ: Hampton Press.

Moss, A., D. Scott, and J. P. Zibin. 2011. "No Evidence for Anomalously Low Variance Circles on the Sky." *Journal of Cosmology and Astroparticle Physics* 2011, 1–7.

Mou, B. 2009. *Routledge History of World Philosophies*. Vol. 3, *History of Chinese Philosophy*. Abingdon, England: Routledge.

Murrell, B. 2022. "The Cosmic Plenum: Teilhard's Gnosis: Cosmogenesis." Stoa del Sol website.

NASA. 2013. "Hubble Finds Birth Certificate of Oldest Known Star." NASA website.

Needham, N. R. 1998. *2000 Years of Christ's Power*. Vol. 1, *The Age of the Early Church Fathers*. London: Grace Publications Trust.

Newton, I. 1846. *Newton's Principia: The Mathematical Principles of Natural*

Philosophy. Translated by A. Motte. New York: Daniel Adee. Available at archive.org. Originally published in Latin by I. Newton and J. Machin. 1729. *The Mathematical Principles of Natural Philosophy*. Vol. 1. Translated by A. Motte. London: Benjamin Motte. Available at hathitrust.org.

Nielsen, B. 2003. *A Companion to Yi Jing Numerology and Cosmology: Chinese Studies of Images and Numbers from Han (202 BCE–220 CE) to Song (960–1279 CE)*. London: RoutledgeCurzon.

Northern Shaolin. 2019. "Five Elements: Cycles of Balance" (graphic). Northern Shaolin Academy website.

Norton, J. D. 2000. "Nature Is the Realisation of the Simplest Conceivable Mathematical Ideas: Einstein and the Canon of Mathematical Simplicity." *Studies in History and Philosophy of Science Part B: Studies in History and Philosophy of Modern Physics* 31 (2): 135–70.

Nylan, M. 2001. *The Five Confucian Classics*. New Haven, CT: Yale University Press.

"Objecthood." 2019. Wiktionary.

Oliphant, M. 1993. *The Earliest Civilizations*. New York: Simon and Schuster.

Pankenier, D. W. 1981. "Astronomical Dates in Shang and Western Zhou." *Early China* 7, 2–37.

———. 1983. "Mozi and the Date of Xia, Shang, and Zhou: A Research Note." *Early China* 9–10, 175–83.

———. 1998. Astrological origins of heaven's mandate and five elements theory. In: *Ancient modes of thinking in China and origin of yin-yang and wuxing doctrine*, edited by Ai Lan, Wang Tao, and Fan Yu-zhou. Nanking: Jiangsu Classics Publishing House.

PBS. 1997. *Mysteries of Deep Space: Interactive Timeline*. New York: Engel Brothers Media/Thomas Lucas Productions. PBS website.

Pedersen, O. 2011. *A Survey of the Almagest: With Annotation and New Commentary by Alexander Jones*. New York, NY: Springer.

Penrose, R. 1969. "Gravitational Collapse: The Role of General Relativity." *Rivista del Nuovo Cimento* 1, 252–76.

———. 1989. *Shadows of the Mind: A Search for the Missing Science of Consciousness*. Oxford: Oxford University Press.

———. 2005. "Before the Big Bang: An Outrageous New Perspective and Its Implications for Particle Physics." In *Proceedings of the 10th European Particle Accelerator Conference 2006*, edited by C. Prior, 2759–62. Geneva, Switzerland: CERN.

———. 2010. *Cycles of Time: An Extraordinary New View of the Universe.* London: The Bodley Head.

———. 2012. “The Basic Ideas of Conformal Cyclic Cosmology.” *Conference Proceedings of the American Institute of Physics* 1446, 233–43.

———. 2014. “On the Gravitization of Quantum Mechanics 2: Conformal Cyclic Cosmology.” *Foundations of Physics* 44 (8): 873–90.

Penrose, R., and S. Hameroff. 2011. “Consciousness in the Universe: Neuroscience, Quantum Space-Time Geometry and Orch OR Theory. *Journal of Cosmology* 14, 1–50.

Penrose, R., S. Hameroff, and H. P. Stapp. 2011. *Consciousness and the Universe: Quantum Physics, Evolution, Brain and Mind.* Cambridge, MA: Cosmology Science Publishers.

Peratt, A. L. 1995. “Introduction to Plasma Astrophysics and Cosmology.” *Astrophysics and Space Science* 227 (1–2), 3–11.

Perkins, F. 2019. “Metaphysics in Chinese Philosophy.” *The Stanford Encyclopedia of Philosophy* website.

Pettinari, G. W. 2016. *The Intrinsic Bispectrum of the Cosmic Microwave Background.* Cham, Switzerland: Springer International.

Phillips, R., J. Kondev, J. Theriot, and H. G. Garcia. 2013. *Physical Biology of the Cell.* 2nd ed. New York: Garland Science.

Planck Collaboration. 2014. “Planck 2013 Results. XVI. Cosmological Parameters.” *Astronomy and Astrophysics* 571, A16.

———. 2016. “Planck 2015 Results: XIII. Cosmological Parameters.” *Astronomy and Astrophysics* 594, A13.

Possehl, G. L. 2002. *The Indus Civilization: A Contemporary Perspective.* Lanham, MD: Rowman AltaMira.

Pregadio, F. 2008. *The Encyclopedia of Daoism.* 2 vols. Abingdon, England: Routledge.

Prokopec, T. 2011. “Negative Energy Cosmology and the Cosmological Constant.” Available on the Archive of Cornell University website (arXiv).

Pultarova, T. 2017. “Does Dark Matter Exist? Bold New Study Offers Alternative Model.” Space.com.

Raghav. 2016. *Motivating Thoughts of Stephen Hawking.* New Delhi: Prabhat Prakashan.

Redmond, G., and T. K. Hon. 2014. *Teaching the I Ching (Book of Changes).* Oxford: Oxford University Press.

Ren, J. Y. 1998. *History of the development of Chinese philosophy*. Beijing: People Press.

Ricci, Matteo. 1985. *The True Meaning of the Lord of Heaven*. Edited by E. J. Malatesta, translated by D. Lancashire and P. K. Hu. Chestnut Hill, MA: Institute of Jesuit Sources.

Richard, T. 1890. "The Influence of Buddhism in China." *Chinese Recorder* 21 (2): 63–64.

Robinet, I. 2008. "Taiji Tu: Diagram of the Great Ultimate." In *The Encyclopedia of Daoism A–Z*, edited by F. Pregadio, 934–36. Abingdon, England: Routledge.

Roland, A. L. 2009. "Growing Scientific Consciousness Revolution." World News Trust website.

Rutt, R. 2002. *Zhouyi: A New Translation with Commentary of the Book of Changes*. London: RoutledgeCurzon.

Sadakata, A. 2009. *Buddhist Cosmology: Philosophy and Origins*. Tokyo: Kosei Publishing.

Sawyer, T. J. 1882. *Endless Punishment: In the Very Words of Its Advocates*. Boston: Universalist Publishing House.

Schirokauer, C., and M. Brown. 2011. *A Brief History of Chinese Civilization*. 4th ed. Boston: Wadsworth Cengage Learning.

Schlesinger, A. 1974. "The Missionary Enterprise and Theories of Imperialism." In *The Missionary Enterprise in China and America*, edited by J. K. Fairbank, 336–75. Cambridge, MA: Harvard University Press.

Schombert, J. 2016a. *Lecture 1: Ancient Cosmology*. Online lecture notes. Astronomy 123: Galaxies and the Expanding Universe, University of Oregon website.

———. 2016b. *Lecture 2: Medieval Cosmology* (Online lecture notes). Astronomy 123: Galaxies and the Expanding Universe, University of Oregon website.

Serway, R. A., and J. W. Jewett. 2014. *Physics for Scientists and Engineers with Modern Physics*. 9th ed. Belmont, CA: Brooks/Cole.

Seshavatharam, U. V. S., and S. Lakshminarayana. 2016. "Is Dark Energy an Alias of Cosmic Rotational Kinetic Energy? *International Journal of Advanced Astronomy* 4 (2): 90–94.

Shao, W. H. 2004. *China Encyclopedia of Fengshui*. Lhasa: Tibetan People Publishing House.

Shao, Y. 1993. *Treatise of supreme world-ordering principles*. Zhengzhou, China: Zhongzhou Ancient Works Publishing House.

———. 2003. *Supreme world-ordering principles*. Beijing, China: Jiu-zhou Press.

———. Song-a. Treatise of supreme world-ordering principles. Chinese Text Project website. Annotated catalog of the complete imperial library.

———. Song-b. "Supreme world-ordering principles 1–14." In: *Right-governance-year Daoist canon in Ming dynasty*. Supreme-mystery part. Chinese Text Project website.

Shukla, P. K., and B. Eliasson. 2010. "Nonlinear Aspects of Quantum Plasma Physics." *Physics-Uspekhi* 53 (1): 51–76.

Sideris, L. H. 2017. *Consecrating Science: Wonder, Knowledge, and the Natural World*. Oakland: University of California Press.

Slocum, T. 2006. Review of *The Powers of the Universe*. In *Presence: An International Journal of Spiritual Direction*.

Smeenk, C. 2013. "Philosophy of Cosmology." In *Oxford Handbook of Philosophy of Physics*, edited by R. Batterman, 607–52. Oxford: Oxford University Press.

Smeenk, C., and G. Ellis. 2017. "Philosophy of Cosmology." *The Stanford Encyclopedia of Philosophy* website.

Smith, B. 2001. "Objects and Their Environments: From Aristotle to Ecological Ontology." In *The Life and Motion of Socioeconomic Units*, GISDATA, vol. 8. Edited by A. Frank, J. Raper, and J.-P. Cheylan, 69–87. London: Taylor and Francis.

Soothill, W. E. 1924. *Timothy Richard of China: Seer, Statesman, Missionary and the Most Disinterested Adviser the Chinese Ever Had*. London: Seeley, Service and Co.

Spence, J. D. 1985. *The Memory Palace of Matteo Ricci*. New York: Penguin Books.

Steinhardt, P. J., and N. Turok. 2002. "A Cyclic Model of the Universe." *Science* 296 (5572): 1436–39.

———. 2005. "The Cyclic Model Simplified." *New Astronomy Reviews* 49 (2–6): 43–57.

Stringer, C. B., and P. Andrews. 1988. "Genetic and Fossil Evidence for the Origin of Modern Humans." *Science* 239 (4845): 1263–68.

Strohman, J. M. 2017. *Application Commentary of the Gospel of Matthew*. Rev. ed. Pierre, SD: Cross Centered Press.

Sung, Z. D. 1973. *The Text of Yi King and Its Appendixes: Chinese Original with English translation*. Taipei: Culture and Books.

Swimme, B. T. 2004. "Announcing a New DVD Series from Cosmologist

Brian Swimme." *Gatherings: Journal of the International Community for Ecopsychology.*

———. 2017a. *Cosmological Powers.* PARP 6110, course syllabus. California Institute of Integral Studies website.

———. 2017b. [Swimme's] Lecture notes on *The Powers of the Universe.* 3 DVD set. Unpublished manuscript.

Swimme, B. T., and C. Busch. 1990. *Canticle to the Cosmos.* DVD. Center for the Story of the Universe website.

Swimme, B. T., and D. Anderson. 2004. *The Powers of the Universe.* 3 DVD set. Center for the Story of the Universe website.

Swimme, B. T., and T. Berry. 1992. *The Universe Story: From the Primordial Flaring Forth to the Ecozoic Era: A Celebration of the Unfolding of the Cosmos.* San Francisco, CA: HarperOne.

Swimme, B. T., and M. E. Tucker. 2011. *Journey of the Universe.* New Haven, CT: Yale University Press.

TBF. n.d. *The Influence of Teilhard.* Thomas Berry Foundation website.

Teall, E. K. 2014. "Medicine and Doctoring in Ancient Mesopotamia." *Grand Valley Journal of History* 3 (1): 1–8.

Teilhard de Chardin, P. 1964. *The Future of Man.* Translated by N. Denny. New York: HarperCollins.

———. 1975a. *The Phenomenon of Man.* Translated by B. Wall. 2nd ed. New York: Harper Colophon.

———. 1975b. *Toward the Future.* Translated by R. Hague. New York: Harcourt Brace Jovanovich.

———. 1999. *The Human Phenomenon: A new Edition and Translation of "Le phenomene humain."* Translated by S. Appleton-Weber. Brighton, England: Sussex Academic Press.

Teiser, S. F. 2002. "The Spirits of Chinese Religion." In *Religions of Asia in Practice: An Anthology.* Edited by D. S. Lopez, 295–329. Princeton, NJ: Princeton University Press.

Teplan, M. 2002. "Fundamentals of EEG Measurement." *Measurement Science Review* 2 (2): 1–11.

Terzić, B. 2017. "Phys 652 Astrophysics: Lecture 13: History of the Very Early Universe." Northern Illinois Center for Accelerator and Detector Development website.

Theuns, T. 2016. *Physical Cosmology* (lecture in PDF). Durham University: Institute for Computational Cosmology website.

"Thingness." 2019. *Merriam-Webster's online dictionary.*

Thomson, I. D. 2005. *Heidegger on Ontotheology: Technology and the Politics of Education*. Cambridge: Cambridge University Press.

Toomer, G. J. 1988. "Hipparchus and Babylonian Astronomy." In *A Scientific Humanist: Studies in Memory of Abraham Sachs*, edited by E. Leichty, M. de J. Ellis, and P. Gerardi, 353–62. Philadelphia, PA: Samuel Noah Kramer Fund, The University Museum.

———. 1996. "Ptolemy and His Greek Predecessors." In *Astronomy Before the Telescope*, edited by C. B. F. Walker, 68–91. London: British Museum Press.

Tuchtey, B. 2011. *Colonialism and Imperialism, 1450–1950*. Europäische Geschichte Online (EGO) website.

Tyshetskiy, Y. O., S. V. Vladimirov, and R. Kompaneets. 2013. "Unusual Physics of Quantum Plasmas." *Voprosy atomnoj nauki i techniki / Pytannja atomnoï nauky i techniky / Problems of atomic science and technology* 83 (1): 76–81.

Udías, A. 2005. "Teilhard de Chardin and the Present Dialogue between Science and Religion." *Pensamiento* 61 (230): 209–29.

———. 2009. *Christogenesis: The Development of Teilhard's Cosmic Christology.* Woodbridge, CT: American Teilhard Association.

———. 2016. "Pierre Teilhard de Chardin: the Spirituality of a Man of Science." *ITEST Bulletin* 47 (3): 3–8.

UNESCO. 2016. "The Twenty-Four Solar Terms: Knowledge in China of Time and Practices Developed through Observation of the Sun's Annual Motion."

Ureview. 2019. "Review of the Universe—Structures, Evolutions, Observations, and Theories." The observable universe and beyond website.

Vale, C. J. 1992. "Teilhard de Chardin: Ontogenesis vs. Ontology." *Theological Studies* 53, 313–37.

Valev, D. 2014. "Estimations of Total Mass and Energy of the Observable Universe." *Physics International* 5 (1): 15–20.

Van Flandern, T. 2002. "The Top 30 Problems with the Big Bang Theory. *Meta Research Bulletin* 11, 6–13.

Vieira, A. 2001. *Chave dos profetas, 3*. Edited by A. E. Santo, translated by J. P. Gomes. Lisbon, Portugal: Biblioteca Nacional Portugal.

Vikoulov, A. 2017. "From the Holographic Principle to the Holofractal Principle." Ecstadelic website.

Von Humboldt, A. 1866. *Cosmos: A Sketch of a Physical Description of the Universe*. Translated by E. C. Otté. New York: Harper and Brothers.

Walls, J., and Y. Walls. 1984. *Classical Chinese Myths*. Hong Kong: Joint Publishing Company.

Wang, D. Y. Song dynasty. "Commentary to the marvelous life-protection scripture of the celestial child of the supreme and utmost purity." In: *Explanations and commentary with diagrams to the wondrous canon of the eternal purity and tranquility as taught by the Supreme Venerable Sovereign*. Chinese Text Project website.

Wang, H., B. Wang, K. P. Normoyle, et al. 2014. "Brain Temperature and Its Fundamental Properties: A Review for Clinical Neuroscientists." *Frontiers in Neuroscience* 8 (307): 1–17.

Wang, L. X. 1997. *American missionaries and the modernization of China in the late Qing*. Tianjin, China: Tianjin People Press.

Wang, Y., J. M. Kratochvil, A. Linde, et al. 2004. "Current Observational Constraints on Cosmic Doomsday." *Journal of Cosmology and Astroparticle Physics* 12 (006): 1–19.

Weinberg, S. 1992. *Dreams of a Final Theory: The Scientist's Search for the Ultimate Laws of Nature*. New York: Pantheon Books.

Welch, H. 1965. *Daoism: The Parting of the Way*. Boston: Beacon Press.

Wetterich, C. 2013. "Universe without Expansion." *Physics of the Dark Universe* 2 (4): 184–87.

Whittmore, T. 1840. *Plain Guide to Universalism*. Boston. Available at archive.org.

Wilber, K. 2001. *A Theory of Everything: An Integral Vision for Business, Politics, Science, and Spirituality*. Boston: Shambhala Publications.

Wilhelm, R. 1990. *The I Ching or Book of Changes*. Translated by C. F. Baynes. Princeton, NJ: Princeton University Press.

Williams, L. P. 2018. "History of Science." *Encyclopædia Britannica* online.

Wright, E. L. 2019. "Frequently Asked Questions in Cosmology: What Is the Evidence for the Big Bang?" UCLA Division of Astronomy and Astrophysics.

Wu, C. Yuan dynasty. *A collective interpretation of the 72 phenological terms*. Chinese Text Project website.

Wyatt, D. J. 2010. "Shao Yong's Numerological-Cosmological System." In *Dao Companion to Neo-Confucian Philosophy*. Edited by J. Makeham, 17–38. Dordrecht, Netherlands: Springer.

Xiao, J. 2014. *Encyclopedia of Wuxing: Full interpretation in vernacular Chinese*. Annotated and translated by H. Y. Liu and B. L. Liu. Beijing, China: China Meteorological Press.

Xu, D. S. 1997. *A history of Daoism*. Shanghai, China: East China Normal University Press.

Xu, S. 2002 (ed.). "On the beginning of Wuxing's births." In: *Abyss Zi-Ping*, Song dynasty. Annotated by Li Feng. Haikou, China: Hai Nan Press.

Yang, F. J. 2008. *Earliest Chinese ancestor, Taihao Fu Xi: Origin of ancient civilization in China, II*. Shanghai, China: Shanghai University Press.

Yang, S. n.d. "Interpretation of Fu-xi 64-hexagram square-circle diagram." wenku website.

York, D. 1997. *In Search of Lost Time*. Boca Raton, FL: CRC Press.

Yu, Y. 2004. *The separate transmission of the book of changes (1284)*. In: *China Tao Tsang*. Edited by J. Y. Zhang and M. C. Liao, 14, 18. Beijing: Huaxia Press.

Zhang, D. N. 1982. *Outline of Chinese philosophy*. Beijing: China Social Science Press.

Zhang, J., and Y. Li. 2001. "'Mutual Stealing among the Three Powers' in the Scripture of Unconscious Unification." In *Daoism and Ecology: Ways within a Cosmic Landscape*, edited by N. J. Girardot, J. Miller, and X. Liu, 113–24. Cambridge, MA: Harvard University Press.

Zhang, J. F. "Northern Song dynasty. Section of primordial Qi, primordial chaos, primordial beginning, and eon phases." Chinese Text Project website. See also: "Seven tablets in a cloudy satchel," 2, 6. Chinese Text Project website.

Zhang, X. C. Southern Song dynasty. *Outer chapters to observe matters of supreme principles to govern the world*. Vol. 8. Chinese Text Project website.

Zheng, J. H. and F. Yue. 2011. "A new survey of Nicolas Trigault's contributions to Sino-Western cultural exchange." *Oriental Forum* 2, 38–43.

Zheng, K. C. n.d. "Chiseling to open the regularities of the hexagram Qian in Zhou-yi." Chinese Text Project website.

Zheng, W. L. 2007. "Tentative study of Zhu Xi's li-qicosmology." *Philosophy and Culture* 34 (3): 1–14.

Zhou, D. Y. 1990. *Collection of Zhou Dun-yi*. Annotated by Chen Ke-ming. Beijing: Zhonghua Book Company.

Zhou, X. 2017. *Code of the Yijing*. Hong Kong: Zhen Yuan Press.

Zhu, Q. 1993. *Nestorian christianity in China*. Beijing: People Publication Press.

Zhu, W. 2007. *Matteo Ricci's complete corpus in Chinese*. Shanghai, China: Fudan University Press.

Zhu, Xi. Song dynasty. *Literary corpus of Master Zhu Xi*. Vol. 58, chapter 26. Chinese Text Project website

———. 1992. *Original meaning of Zhou-I.* Annotated by Su Yong. Beijing, China: Peking University Press.

Zhuangzi. 350–250 BCE. Chapter 5 of *knowledge rambling in the north.* In *Outer chapters of Zhuang-Zi.* Translated by J. Legge. Chinese Text Project website.

———. Chapter 11 of *Geng-sang Chu.* In *Miscellaneous chapters of Zhuang-Zi.* Chinese Text Project website.

Zinkernagel, H. 2014. "Philosophical Aspects of Modern Cosmology." *Studies in History and Philosophy of Modern Physics* 46. Special issue on philosophy of cosmology, 1–4.

Zorba, S. 2012. *Dark Energy and Dark Matter as Inertial Effects.* Available on the Archive of Cornell University website (arXiv).

Zurich, E. T. H. 2013. "Why an Ice Age Occurs Every 100,000 Years: Climate and Feedback Effects Explained." *ScienceDaily* website.

Index